# Harmful Natural Chemicals and Radiation in the Environment

Stories, History and
What You Need to Know

# Harmful Natural Chemicals and Radiation in the Environment

## Stories, History and What You Need to Know

**Raymond Poon** *Ph.D.*

**World Scientific**

NEW JERSEY • LONDON • SINGAPORE • BEIJING • SHANGHAI • HONG KONG • TAIPEI • CHENNAI

*Published by*

World Scientific Publishing Co. Pte. Ltd.

5 Toh Tuck Link, Singapore 596224

*USA office:* 27 Warren Street, Suite 401-402, Hackensack, NJ 07601

*UK office:* 57 Shelton Street, Covent Garden, London WC2H 9HE

**British Library Cataloguing-in-Publication Data**
A catalogue record for this book is available from the British Library.

ISBN 978-981-4412-93-3

Typeset by Stallion Press
Email: enquiries@stallionpress.com

Printed by FuIsland Offset Printing (S) Pte Ltd Singapore

To my dear wife Anita, and my daughters,
Carol, Joanne, Grace and Lillian.

# ACKNOWLEDGMENT

The author wishes to thank Dr. Jason Chan for reviewing the manuscript.

# PREFACE

Synthetic pollutants have occupied public attention for a long time. This is particularly true of synthetic pollutants such as pesticides, polychlorinated biphenyls (PCBs) and dioxins which emerged in the past several decades. These pollutants are of interest to us because they are persistent in the environment, prevalent in our everyday life, and, to various extents, hazardous to our health. For these reasons much research has been conducted on them and books and articles have been written about their environmental, ecological and health effects. In contrast, naturally occurring chemicals and radiations receive much less consideration even though they are no less persistent, prevalent and toxic. In fact, the impact of naturally occurring chemicals and radiations on our daily life and health far exceeded that of many synthetic pollutants. Their roles in mass poisoning, disastrous events and public health problems are well-documented if not well-remembered. For example, arsenic in well water maimed and killed hundreds of thousands of people worldwide in the past four decades alone and remains a real threat today; carbon dioxide released from lake turnovers took the life of almost the entire population of nearby villages; and lead in drinking water and dusts remain a constant public health concern today. One must not ignore the fact that natural radiations, both radioactive and non-radioactive, can be extremely persistent and harmful to health. We are constantly exposed to radon, a radioactive gas, be it indoor or outdoor. Nor can we avoid the effect of heat which is a form of nonradioactive radiation. Heat waves, in fact, have and will continue to kill tens of thousands whenever and wherever they strike.

There is a need to balance our perception of the risk of synthetic pollutants versus naturally occurring chemicals and radiation. A primary purpose of this book is to describe, in a narrative form, the hazards of naturally occurring chemicals and radiation, with the hope that the information will help redress this imbalance. It is also hoped that the information will prompt the scientific and regulatory communities to focus their effort on understanding, preventing, and mitigating the harmful effects of naturally occurring chemicals and radiations.

Another motivation for writing this book is a concern about verifiability of information. In the age of free and instant internet communication, the term "The medium is the message"[1] comes alive. Any message that is written in a crisp contemporary language, appropriately "branded", and conveyed through popular electronic/ digital media, will capture the attention of readers. Furthermore, readers may be sufficiently impressed by the packaging of such messages that they accept them as facts without bothering to verify the content or check the background of original senders. Let me illustrate with an example. I received a professionally appearing email, unsigned of course, touting the benefit of almond as a cancer fighting food. To bolster the claim it cited the "fact" that Fijians are mostly cancer-free because almond is part of their staple diet. There is indeed some preliminary experimental evidence that ingredients in almond may have antitumor activity.[2,3] However, the cancer-free claim was patently false. Fijians suffer from all forms of cancers,[4] and suffer from a high rate of cervical cancer.[5] I called this a "chimeric" message, not only because of its mythical and fanciful component, but also the dangerously misleading combination of facts and fiction. Following the tradition of a science book, I have referenced all toxicity data and health information. Furthermore, the best effort has been made to reference all histories, stories, anecdotes and quotes so that they are traceable, if not completely verifiable. This does not mean the book is error-free, but any mistake is in black and white, traceable to the author and authors of the cited articles, and, mostly importantly, open to criticism and amenable to revisions and corrections.

A book of this nature need not be boring. Men have been living with chemicals and radiations as long as they existed on Earth and the ancient and modern history of their interactions are most interesting and revealing. There is also a treasure trove of stories and anecdotes accompanying these chemical and radiation that serve to illustrate their deep impact on society and our health and well-being. Stories of men's successes and failures in dealing with these hazards are plenty.

This book should be useful as an introductory textbook for students in senior high schools and universities who are interested in environmental studies. It may also be of help to undergraduates who wish to pursue further study or research in the field of environmental toxicology. Environmental engineers, teachers and scientists in the field of environmental toxicology may find the book informative because of its comprehensive and referenced coverage of important natural contaminants and radiations. Urban planners, health officials, policy makers and regulators may be able to use the book as a source of information and reference. The science and toxicological components are presented as much as possible in layman's language, and interwoven with relevant history and stories. Hopefully the general public will find this book an interesting read.

## References

1. Marshall McLuhan. (2001) Understanding Media: The Extension of Man. Routledge, London and New York.
2. Mandalari G, Tomaino A, Arcoraci T *et al.* (2010) Characterization of polyphenols, lipids and dietary fibre from skins of almonds (Amygdalus communis L.). *J. Food Comp. Anal.* **23**: 166–174.
3. Liu RH. (2004) Potential synergy of phytochemicals in cancer prevention: mechanism of action. *J. Nutr.* **134**: 3479S–3485S.
4. Moore MA, Baumann F, Foliaki S *et al.* (2010) Cancer epidemiology in the pacific islands — past, present and future. *Asian Pac. J. Cancer Prev.* **11**, Suppl 2: 99–106.
5. HPV and Cervical Cancer in the World. Vaccine. 2007 Report Volume 25, Supplement 3, 1 November 2007. [http://www.who.int/hpvcentre/publications/HPVReport2007.pdf].

# CONTENTS

# INTRODUCTION

The basic concept of risk assessment that is embraced by the scientific and regulatory communities is: RISK = HAZARD × EXPOSURE.[1] Within the context of this book a hazard is the property of a naturally occurring chemical or radiation that causes damages, harms or adverse health effects, and exposure is the contact, or opportunity of contact, with a potentially hazardous source by various routes over a period of time. Risk can thus be defined as the probability of experiencing harms or adverse health effects following exposure to a hazard over a defined period of time. In this book, the primary focus is on human hazards (epidemiological evidence, disease evidences, case reports), followed by experimentally generated toxicity data from animal, cellular, biochemical and molecular studies.

Understandably the book is focused on high risk chemicals and radiations. Arsenic, for example, is a chemical of significant risk because it is capable of causing dermal lesions, skin cancer, and a variety of other adverse effects, and because millions of people are still expose to it through drinking water, food and dust. Aluminum, on the other hand, has not been irrefutably established to be highly toxic to humans. However, it poses a risk because of its ubiquitous presence in our everyday life (high exposure) and, therefore, merits attention. An exception is ethylene — a low- hazard and low-exposure chemical. A chapter was nevertheless devoted to ethylene because it has interesting and instructive stories to tell which involved history, mythology, archeology, geology, chemistry and toxicology.

Our appreciation of the adverse effects of radiation is often hampered by confusion about the meaning of radiation and the myriad of units of measurement used. A chapter was therefore devoted to the fundamental and history of radiation and radio-activity. One of the focuses of the book is to emphasize the fact that heat and UV-radiation are essential parts of naturally occurring environmental hazards. This inclusion will help us gain a comprehensive view of the relative contribution of each and every environmental hazard and hence achieve a realistic assessment of their relative contribution to risk.

All chapters were not formatted the same way because each topic has a different emphasis, be it environmental/global distribution, risk characterizations, case reports, health impacts, or historical perspectives. The chapter on copper, for example, followed the traditional section layout of occurrence and use, history, and health effects. On the other hand the chapter on toxic gases consisted of three different gases and each with its own narrative format. Readers may find it difficult to find a common theme because of the varied formats. However, they will be able to find in many chapters common example of how the lack of foresight, wisdom and prudence on the part of scientists, regulators and the industry led to confusions, mistakes and even disasters. All chapters have in common a section on Practical Points. It should be noted that this section is in the nature of a summary of experiences and advices from experts, and is not meant to be a definitive or an authoritative guide.

Whenever possible, data and information published in books and peer-reviewed journals are used. Many regulatory bodies such as the U.S. Agency for Toxic Substances and Disease Registry and World Health Organization disseminate information online and their websites are cited (dated if possible). In a few cases it is necessary to reference media websites that may be transient.

## Reference

1. NRC (National Research Council). (1983) *Risk Assessment in the Federal Government: Managing the Process.* Washington D.C.: National Academy Press.

# CHAPTER 1

# ALUMINUM

## Aluminum in Nature

Aluminum is a silvery-white, malleable and ductile metal in Group 13 (III) of the periodic table. Aluminum is the third most abundant element (after oxygen and silicon), and the most abundant metal, making up about 8% by weight of the Earth's solid surface. Aluminum in its elemental form is not found in nature because it is highly reactive and is rapidly oxidized to oxides or combined with other element to form aluminum compounds. Aluminum compounds are found everywhere. Clay, for example, is composed of aluminum, silicon and oxygen with trace amount of other metals. Feldspar is aluminum silicate found in igneous rocks. Corundum, a mineral next only to diamond in its hardness, is a form of aluminum oxide. Bauxite, which contains up to 55% aluminum oxide (alumina), is the major raw material for the extraction of aluminum. Some gemstones are aluminum oxides whose specific color is determined by the type of trace metals they contain, e.g. chromium imparts the characteristic red color to ruby, while the violet color of sapphire is due to iron and titanium.

## Use of Aluminum

Aluminum metal is used in beverage cans, cooking utensils, tin foils and coins. Alloys of aluminum are strong, light-weight, easy to shape and weld, and corrosive-resistant. Therefore, aluminum is extensively used in the car and aircraft industries, in the building trades, and in decorative arts and jewelry. Aluminum is an ingredient in some solid fuels, rocket propellants and explosives. Interestingly, aluminum

1

compounds are also used as fire retardants and as foaming agents in fire extinguishers. Because of its excellent reflective and low radiant properties, aluminum foils and alloys are extensively exploited in spacecraft and space suits. Aluminum is used in power lines, electrical conductors, cables and wiring. Aluminum sulfate (alum) is used as mordant in the dye industry and in large quantity as coagulant in water treatment. Corundum is second in hardness to diamond and an important industrial abrasive. Aluminum oxide is employed in laboratories and industies for separating mixtures. Some of the aluminum-containing products are in close contact with our body, e.g. jewelry, dental crowns and denture materials, cosmetics and antiperspirants. A wide variety of aluminum products are food additives; some are foaming, emulsifying and anti-caking agents, while others are added for the purpose of pH and color adjustment. Aluminum may be present in various concentration in medicinal products such as antacid, buffered aspirins, and adjuvant for vaccines.

From the early 1940s to late 1970s, inhalation of aluminum particles was experimented as a therapy for silicosis,[75,76] a worldwide occupational lung disease caused by breathing in air contaminated by crystalline silicon dioxide or silica.[77] The rationale for such a therapeutic procedure was that aluminum can coat the silica particles thus reducing their solubility and reactivity toward lung tissues. For a time, miners in Australia, the United States and Canada liable to silica dust exposure were sprayed with an aluminum mist before they entered the work area. However, the effectiveness of aluminum therapy for the prevention or amelioration of silicosis has not been unequivocally demonstrated.[77,88] Furthermore, concern was raised that the experimental procedure undermined the importance of primary preventive measure, namely the reduction of silica dust in workplace air.[78]

## History

Although aluminum compounds are as old as earth itself, the elemental metallic form was not known to the world until the 19th century. In 1825, the Danish scientist Hans Christian Ørsted succeeded in isolating a reasonably pure form of elemental aluminum by reacting anhydrous aluminum oxides with a reactive potassium mixture.[1] Aluminum is tightly bound to other elements such as oxygen and it takes tremendous amounts of energy to separate these bonds and extract the elemental metal. The separation process was time-consuming and costly, and the aluminum produced was considered a precious metal; a major use was to be fashioned into jewelry items. In 1886, the American Charles Martin Hall and the Frenchman Paul Héroult separately invented the electrolytic reduction process for the production of aluminum from alumina extracted from bauxite.[1] Although the Hall–Héroult process made large scale production possible, aluminum was still an expensive metal because, at that time, the cost of electricity required for the electrolytic process was high. For example, the aluminum block used to cap the Washington Monument in 1884 costs $16 per pound, the same as the then prevailing market price of silver.[2] It was not until the turn of the 20th century, when the generation of electricity becomes less costly, that widespread application of aluminum in fabrication, industry and consumer goods began. For this reason aluminum is often called the metal of the 20th century.

## Aluminum in Air, Water and Food

*Air.* Aluminum found in the atmosphere is predominantly in particulate forms. Because aluminum accounts for 8% of the Earth's crust, much of the atmospheric source of aluminum originates from natural erosion of soil and rocks and volcanic activity.[3] Anthropogenic sources account for about 13% of atmospheric aluminum[4] and these include aluminum production, coal combustion, exhaust from motor vehicle combustion, mining and agricultural activities, and other industrial activities. Airborne aluminum levels range from 0.0005 $\mu g/m^3$ over Antarctica to more than 1 $\mu g/m^3$ in industrialized areas.[5]

***Water.*** Aluminum occurs ubiquitously in natural waters as a result of the weathering of aluminum-containing rocks and minerals. Atmospheric deposition is another natural source of aluminum input to surface water. Anthropogenic sources such as mining activity, industrial and municipal discharges, surface run-off, may also contribute to aluminum in surface water. Wet and dry deposition of atmospheric aluminum to surface water is a constant input.[3,5] Low pH facilitates mobilization and solubility of aluminum; acidic mine drainage and acid rain for example, cause an increase in the dissolved aluminum content of surrounding waters.[6,7] Aluminum concentration in drinking water varies widely from country to country and from region to region. In an analysis of drinking water from New York City and 25 states in the U.S., the aluminum concentrations were found to range from undetectable to 1.029 mg/L.[8] In the city of Xi'an, China, total aluminum concentrations in drinking water were found to range from 0.051–0.417 mg/L.[9] In the Kirazli region of Turkey,[10] aluminum concentration as high as 15 mg/L has been reported in water that may be used for drinking purpose. Treatment of drinking water with alum or other aluminum-based coagulant for water treatment may contribute to the aluminum concentration in drinking water.[3,9] Normally, the contribution of drinking-water to the total oral exposure to aluminum is about 4%.[5] The World Health Organization does not set a limit for aluminum on drinking water. Instead, it has derived "practicable levels" of ≤0.1 and ≤0.2 mg/L for large and small facilities, respectively, based on optimization of the coagulation process in water treatment plants that use aluminum-based coagulants.[11]

***Food.*** Food is the main source of aluminum intake.[12] Foods such as spinach, tea and some herbs contain high concentrations of aluminum. Some foods with high amounts of aluminum-based additives (e.g. processed cheese, grain products, cream substitutes, and grain-based desserts) contribute significantly to the amount of aluminum we ingest. Cooking with aluminum utensils and drinking beverages from aluminum containers may also add to our oral intake of aluminum. Adult dietary intake of aluminum in industrialized countries ranged from approximately 1.9–11 mg/day. Infant intakes of aluminum in Canada, the United Kingdom, and the USA ranged from 0.03–0.7 mg/day.[5,12]

## Potential Hazards and Health Effect

As much as aluminum is a 20th century metal, aluminum-related diseases are justifiably being called 20th century diseases. Indeed, concerns about known and potential adverse effects related to aluminum continue with the widespread use of aluminum in the 20th century and beyond. The followings are some representative cases that provide insight into the real and putative hazards of aluminum.

*Dialysis Encephalopathy.* Between late 1960s and the 1980s, many kidney disease patients undergoing dialysis came down with a disease called dialysis encephalopathy[13] or dialysis dementia.[14] According to George Dunea,[15] the disease appeared so suddenly that "*it was mysterious and frightening indeed. Patients coming to dialysis would look at each other in fear and wonder whose turn would come next. Consumer groups came to the hospital wondering what was going on. At first, we had no idea what was happening...*". An association of aluminum with dialysis encephalopathy was first reported in 1976 when dialysis patients who developed encephalopathy were found to have higher aluminum concentration in the brain gray-matter. The source of the aluminum was likely from the aluminum-containing phosphate binding gel that these patients were taking orally to control the serum phosphate level.[16] The weakened excretory capability of the kidneys in dialysis patients[17] further exacerbated the accumulation of aluminum. Allen Alfrey[18] considered aluminum intoxication as one of the first of the "new diseases" as a result of advent of chronic dialysis therapy. Later on, it was found that when patients were put on a combined therapy of aluminum-containing phosphate binder and Shohl's solution (a buffer of citric acid and sodium citrate) to prevent hyperphosphatemia and acidosis, the serum aluminum was markedly increased. This was due to citrate enhancing the absorption of aluminum into the body.[19,20] It was also pointed out that aluminum toxicity was manifested in some uremic children who did not undergo dialysis. Again, the cause may be attributed to the concomitant administration of aluminum-containing phosphate binders and sodium citrate for treatment of uremic acidosis.[20] Besides neurological effects, high aluminum also caused bone diseases in dialysis patients.[21] The use of aluminum binders is now limited to short-term management and special cases, and when used, the serum aluminum level is monitored closely.[22]

Another common source of aluminum exposure for dialysis patients is the water with which the dialysates are prepared. Tap water was used in the early days to prepare dialysis fluids. A nationwide epidemiological survey conducted in the UK showed that among 1,293 patients undergoing hemodialysis, the incidence of dialysis encephalopathy and fracturing osteodystrophy was significantly correlated with the aluminum content of the water supply.[23] Ward *et al.*[24] proposed that the best water-treatment method for the preparation of dialysates involved both reverse osmosis and deionization. Today, sophisticated water purification system consisted of prefilters, softeners and reverse osmosis modules is routinely used resulting in reduction in incidence of aluminum intoxication.[25] For example, improvements in water treatment methods and use of non-aluminum-containing phosphate binders in a Philadelphia dialysis center have resulted in low prevalence (<1%) of aluminum toxicity among hemodialysis patients.[26]

A third source of aluminum exposure is through contact with aluminum components and parts in the dialysis system. The occurrence is rare but nevertheless illustrates an un-anticipated negative consequence of incorporating aluminum and its alloy in medical devices. In one incident reported by a dialysis center in Pennsylvania, 64 patients were found to have high serum aluminum; some of these patients experienced seizures and mental status changes requiring hospitalization. It was found that the two pumps that were used to deliver dialysis concentrate to the machines had aluminum components in the casing and casing cover, and the acidity of the concentrate interacted with the aluminum components producing dialysis fluid with high aluminum concentration.[27] In 2001, 27 patients in a Curacao dialysis center were found to have aluminum intoxication; of these 10 died from acute encephalopathy and 17 had minor symptoms and survived. The aluminum was later found to have originated from the aluminum pipe lined on the inside with cement mortar that delivered distilled sea water to the dialysis center. The distilled water corroded the cement mortar and the aluminum leached out and reached the dialysates.[28]

**Aluminum and Alzheimer's Disease.** Alzheimer's disease (AD) is a progressive neurodegenerative disorder and the most common cause of dementia, affecting millions of men and women worldwide. It is characterized by the accumulation of extracellular amyloid-$\beta$ (A$\beta$) plaques and neurofibrillary tangles inside neurons and dystrophic neurons.[29] The connection between aluminum and neurodegenerative diseases was made in 1965 when Klazo *et al.*[30] made an astute observation that intracerebral injection of Holt's adjuvant, which contain alum sulfate as the main ingredient, produced neurofibrillar degeneration in rabbits. Robert Terry and Carlos Pena[31] followed up on the observation and demonstrated by electron microscopy that the neuronal lesion induced by alum sulfate is fibrillar in nature and is similar to the neurofibrillary change observed in human AD. Subsequent studies using sophisticated instrumental analysis showed that the aluminum is located within the neurofibrillary tangle itself.[32] While much data has been accumulated on the physical presence of aluminum in the brain of AD patients, very little is known about the role aluminum plays in the etiology of AD. Much controversy and debate persists today as to whether aluminum is a causative or sole agent for AD.[29,32–34] Current view favors a role of aluminum in the progress and development of AD, e.g. as a cross-linker in $\beta$-amyloid oligomerization, or in the promotion of inflammatory reaction in the brain.[29,34,35] With this picture in mind, we can look at the epidemiological evidence of the association between drinking water aluminum and AD.

Interest in the association of aluminum in drinking water and AD begins with some epidemiological data from Europe. In

England and Wales, Martyn[36] conducted a survey of the rate of AD in people under the age of 70 and reported that the risk of AD was 1.5 times higher in districts where the mean aluminum concentration exceeded 0.11 mg/L as compared to districts where concentrations were less than 0.01 mg/L. Flaten *et al.*[37] studied mortality rate due to dementia in Norway and found that age-adjusted mortality rates grouped by aluminum in drinking water of 0.05 (control), 0.05–0.20, and >0.20 mg/L, showed relative risks for dementia of 1.00, 1.15, and 1.32 in men, and 1.00, 1.19, and 1.42 in women, indicating a dose-response relationship with aluminum concentration in drinking water.[37,38] Neri *et al.*[39] studied 2,344 patients aged 55 or over who, during 1986, had a diagnosis of AD or pre-senile dementia, together with 2,232 controls matched for age and sex. The relative risk was found to be increased with the concentration of aluminum in the drinking water. By 2001, 13 epidemiological studies on aluminum in drinking water and AD had been published with nine out of the 13 reporting statistically significant positive relationship.[38] For 15 years, Rondeau and co-workers[40] followed a group of 1,925 people who were 65 years and older and who were free of dementia at the start and found that daily intake of aluminum was associated with increased risk of dementia and with cognitive decline with time.

Several reviews have assessed the risk posed by aluminum in drinking water and their conclusions are summarized below:

World Health Organization[11]

- The positive relationship between aluminum in drinking-water and AD, which was demonstrated in several epidemiological studies, cannot be totally dismissed.
- The relative risks for AD from exposure to aluminum in drinking-water above 0.1 mg/L, are low (less than 2.0).

U.S. Agency for Toxic Substances and Disease Registry[3]

- The available data do not suggest that aluminum is a causative agent of AD; however, it is possible that it may play a role in disease development.

Krewski *et al.*[41]

- Further studies are needed to settle the debate over the link between aluminum in drinking water and neurological disorders and cognitive impairment.

***The Camelford Incidence.*** In July 1988, a truck driver unloaded 20 tons of aluminum sulfate in the wrong tank at the unmanned water treatment works owned by South West Water Work at Lowermoor, Camelford, North Cornwall. Shortly after the accident, the drinking water became acidic and contaminated with excessive quantities of aluminum, copper, lead, and zinc. Residents of Camelford were exposed to this contaminated water for at least a few days. An advisory group, the Lowermoor Incident Health Advisory Group (LIHAG)[42] was set up in January the following year to investigate the health implications on the population of the Camelford area. *In 1989, The LIHAG issued a report and one of the conclusions is that "...it was not possible to attribute the toxic effects of the incident except insofar as they are a consequence of the sustained anxiety naturally felt by many people".* In October 1990, following representations from the local community, some of whom continued to attribute health problem to the incident, LIHAG was reconvened. A second LIHAG report[43] was issued in November 1991 and part of the conclusion stated that *"The research reported to us does not provide convincing evidence that harmful accumulation of aluminum has occurred, nor that there is a greater prevalence of organic abnormalities in the exposed population. We do not expect lasting physical harm from the toxicity of the contaminated water itself ... Nevertheless, the incident was unique, and the actual doses of aluminum and other contaminants received by the residents are unknown; therefore, although we have no reason to predict any late consequences, we cannot exclude them completely."* In 2001, in response to representations from members of the local community that the health consequences of the incident had not been properly addressed, Health and Environment Ministers asked the Committee on Toxicity of Chemicals in Food, Consumer Products and the Environment (COT), to set up a subcommittee (COT subcommittee) to advise on whether the pollution incident had resulted in delayed or persistent health effects, and on the need for additional monitoring and research (COT). In the meantime several studies on Camelford residents were published relating to aluminum in bones, joint pain, psychological effects, hospital discharge record, pregnancy outcome, and mortality.[44–51] In 1999, Altmann[52] conducted clinical and psychological tests to determine medical condition and anxiety levels in 55 affected subjects. Fifteen siblings who had not been exposed to the contaminated water were used as control. They concluded that people who were exposed to the contaminated water at Camelford suffered considerable damage to cerebral function, which was not related to anxiety. There were a number of reservations about the study due to bias inherent in self selection of cases, flaw in study design and methodology.[53–55] Concern was also raised that the results of the study were obtained under the instruction of plaintiffs' solicitors in the earlier round of litigation.[56] In 2006, Exley and Esiri[57] described a case of a woman who was exposed to the contaminated water in 1988 and died of an unspecified neurological condition fifteen years later. Postmortem examination detected a rare form of abnormal $\beta$ amyloid deposition in her brain. In addition, high concentrations of aluminum were found coincident with the severely affected regions of the cortex. Publication of the long awaited COT subcommittee report, however, is postponed because of an inquest by the West Somerset coroner into the death of an individual who lived in the Lowermoor water supply area at the time of the contamination incident.[43]

The Camelford chronology demonstrates the intense and persistent public interest on aluminum and our health. After half a century of government action and inaction, scientific investigations, inquiries, litigations, public meetings, coroner's inquest, and media attentions, the end is still not in sight. The amount of effort and resource already invested however, may not be a total loss. The experience accumulated is instructive for the risk communication and emergency management communities, not only for aluminum, but for all aspects of drinking water safety.[58]

*Parenteral Nutritional Products.* Aluminum has no known physiological function in humans. An average adult consumes anywhere from 1.9–11 mg per day of aluminum through their diet but it will not accumulate in the body as long as their kidneys are functioning normally. The amount of ingested aluminum that is absorbed by the gastrointestinal tract is estimated to be less than 0.5% so that under ordinary circumstances a person is protected against accumulation of aluminum.[5,12,41,59] As discussed above, it is well-known that when large amounts of aluminum enters our body through dialysis, it can cause adverse effects such as neurotoxicity and bone abnormality. It is therefore not surprising that concern arises when Klein and co-workers[60] reported that patients on long-term total parenteral nutrition had elevated aluminum levels in bone, urine, and plasma, and that the source of aluminum could be traced to the total parenteral solution. They further reported that aluminum may be associated with the development of parenteral nutrition-associated cholestasis in infants.[61] Parenteral nutrition-associated neurotoxicity and metabolic bone effects were also reported and aluminum was attributed to be the cause.[62–64] Although the cholestasis association has not been fully elucidated[41,65] and the other effects needs further corroborative evidence, the US FDA[66] acted by requiring manufacturers of parenteral nutrition solutions to state on their product labels that the product contains no more than 25 mcg/L of aluminum.

*Antacids.* The biggest source of aluminum ingested as a non-food item is aluminum-containing antacids. It has been estimated that the amount of aluminum ingested as antacid is approximately 7,200 mg/day as compared to 1.9–11 mg/day normally consumed as food.[5,12,41] Surprisingly, only a few studies have been conducted on the health effects of antacids. Graves *et al.*[67] reported a positive association between lifetime antacid use and risk of AD. However, this association vanished when consideration was limited to antacids containing aluminum. Osteomalacia has been reported to occur in patients with long history of antacid use[68,69] and in a patient who has taken a large quantity of antacids.[70] Concerns have been raised that concurrent ingestion of antacids with acidic beverages such as those containing citrate may cause a marked augmentation of intestinal aluminum absorption. There is indeed a need for further epidemiological studies on the potential health effects of aluminum-containing antacids.[41]

*Adjuvants.* Aluminum sulfate has been incorporated as an adjuvant in vaccines since 1926.[71] There are documented cases of adverse reaction and contact allergic reaction following injection of vaccines containing aluminum.[41] Aluminum containing adjuvants have been linked with autism,[74] Crohn's disease,[73] and a number of other autoimmune and inflammatory conditions.[72] More studies are needed to evaluate these claims.

## Practical Points

- If you suspect excessive aluminum exposure, your physician can request a blood test or urine test from a certified laboratory; the values from the test will indicate if you have abnormally high concentrations of aluminum.
- Eating large amounts of processed food containing aluminum additives or frequently cooking acidic foods in aluminum pots may expose a person to higher levels of aluminum.[3]
- Limit intake of large quantities of aluminum-containing antacids and buffered aspirin and use these medications only as directed by physicians.[3]
- The major population at risk for aluminum loading and toxicity are individuals with renal failure. Preterm infants may also be particularly sensitive to the toxicity of aluminum due to reduced renal capacity.[3]

# References

1. Knapp B. (1996) *Aluminium.* Atlantic Europe Publishing, Oxon, UK.
2. Binczewski GJ. (1995) The point of a monument: A history of the aluminum cap of the Washington monument. *J. Metals* **47**: 20–25.
3. U.S. Department of Health and Human Services. (2008) Toxicological Profile for Aluminum. Public Health Service. Agency for Toxic Substances and Disease Registry. Atlanta, Georgia.
4. Lantzy RJ, MacKenzie FT. (1997) Atmospheric trace metals: Global cycles and assessment of man's impact. *Geochim. Cosmochim. Acta.* **43**: 511–525.
5. WHO (1998). Aluminium in Drinking-water. Background document for development of WHO Guidelines for Drinking-water Quality. WHO/SDE/WSH/03.04/53. World Health Organization 2003.
6. Davies H, Weber P, Lindsay P *et al.* (2011) Characterization of acid mine drainage in a high rainfall mountain environment, New Zealand. *Sci. Total. Environ.* **409**: 2971–2980.
7. Lawrence GB, Sutherland JW, Boylen CW *et al.* (2007) Acid rain effects on aluminum mobilization clarified by inclusion of strong organic acids. *Environ. Sci. Technol.* **41**: 93–98.
8. Schenk RU, Bjorksten J, Yeager L. (1989) Composition and consequences of aluminum in water, beverages and other ingestibles. In: Lewis TE, ed. Environmental chemistry and toxicology of aluminum. Chelsea, MI: Lewis Publishers, Inc., 247–269.
9. Wang W, Li H, Wang S *et al.* (2010) Spatial variations of aluminum species in drinking water supplies in Xi'an studied applying geographic information system. *J. Environ. Sci.* **22**: 519–525.
10. Bakar C, Karaman HI, Baba A *et al.* (2010) Effect of high aluminum concentration in water resources on human health, case study: Biga Peninsula, northwest part of Turkey. *Arch. Environ. Contam. Toxicol.* **58**: 935–944.
11. WHO (2011). *Guidelines for Drinking-water Quality,* 4th ed. World Health Organization.
12. Soni MG, White SM, Flamm WG *et al.* (2001) Safety evaluation of dietary aluminum. *Regul. Toxicol. Pharmacol.* **33**: 66–79.

13. Alfrey AC, Mishell JM, Burks J *et al.* (1972) Syndrome of dyspraxia and multifocal seizures associated with chronic hemodialysis. *ASAIO* **18**: 257–261.

14. Mahurkar SD, Salta R, Smith EC *et al.* (1973) Dialysis dementia. *Lancet* **1**: 1412–1415.

15. Dunea G. (2001) Dialysis dementia: An epidemic that came and went. *ASAIO* **47**: 192–194.

16. Alfrey AC, LeGendre GR, Kaehny WD. (1976) The dialysis encephalopathy syndrome. Possible aluminum intoxication. *N. Engl. J. Med.* **294**: 184–188.

17. Wills MR, Savory J. (1989) Aluminum and chronic renal failure: Sources, absorption, transport, and toxicity. *Crit. Rev. Clin. Lab. Sci.* **27**: 59–107.

18. Alfrey AC. (1984) Aluminum intoxication. *N. Engl. J. Med.* **310**: 1113–1115.

19. Bakir AA, Hryhorczuk DO, Ahmed S *et al.* (1989) Hyperaluminemia in renal failure: The influence of age and citrate intake. *Clin. Nephrol.* **31**: 40–44.

20. Molitoris BA, Froment DH, Mackenzie TA *et al.* (1989) Citrate: A major factor in the toxicity of orally administered aluminum compounds. *Kidney. Int.* **36**: 949–953.

21. Ihle BU, Becker GJ, Kincaid-Smith P. (1986) Clinical and biochemical features of aluminum-related bone disease. *Kidney. Int.* **29**: S80–S86.

22. Hutchison AJ. (2009) Oral phosphate binders. *Kidney. Int.* **75**: 906–914.

23. Parkinson IS, Ward MK, Feest TG, *et al.* (1979) Fracturing dialysis osteodystrophy and dialysis encephalopathy: An epidemiological survey. *Lancet* **1**: 406–409.

24. Ward MK, Feest TG, Ellis HA *et al.* (1978) Osteomalacic dialysis osteodystrophy: Evidence for a water-borne aetiological agent, probably aluminium. *Lancet* 1841–845.

25. Ledebo I (2009). Purification of dialysis fluid: Historical background and perspective. Blood Purif. **27,** Suppl 1: 17–19.

26. Jaffe JA, Liftman C, Glickman JD (2005). Frequency of elevated serum aluminum levels in adult dialysis patients. *Am. J. Kidney Dis.* **46**: 316–319.

27. Burwen DR, Olsen SM, Bland LA *et al.* (1995) Epidemic aluminum intoxication in hemodialysis patients traced to use of an aluminum pump. *Kidney Int.*: 48469–48474.

28. Berend K, van der Voet G, Boer WH. (2001) Acute aluminum encephalopathy in a dialysis center caused by a cement mortar water distribution pipe. *Kidney. Int.* **59**: 746–753.

29. Shcherbatykh I, Carpenter DO. (2007) The role of metals in the etiology of Alzheimer's disease. *J. Alzheimers. Dis.* **11**: 191–205.

30. Klatzo I, Wisniewski H, Streicher E. (1965) Experimental production of neurofibrillary pathology: 1. Light microscopic observations. *J. Neuropathol. Exp. Neurol.* **24**: 187–199.

31. Terry RD, Pena C. (1965) Experimental production of neurofibrillary degeneration pathology: 2. Electron microscopy, phosphate histochemistry and electron probe analysis. *J. Neuropathol. Exp. Neurol.* **24**: 200–210.

32. Perl DP, Moalem S. (2006) Aluminum and Alzheimer's disease, a personal perspective after 25 years. *J. Alzheimers. Dis.*; 9 (3 Suppl): 291–300.

33. Savory J, Exley C, Forbes WF *et al.* (1996) Can the controversy of the role of aluminum in Alzheimer's disease be resolved? What are the suggested approaches to this controversy and methodological issues to be considered? *J. Toxicol. Environ. Health.* **48**: 615–635.

34. Gupta VB, Anitha S, Hegde ML *et al.* (2005) Aluminium in Alzheimer's disease: Are we still at a crossroad? *Cell. Mol. Life. Sci.* **62**: 143–58.

35. Bondy SC. (2010) The neurotoxicity of environmental aluminum is still an issue. *J. NeuroToxicol.* **31**: 575–581.

36. Martyn CN, Barker DJ, Osmond C *et al.* (1989) Geographical relation between Alzheimer's disease and aluminum in drinking water. *Lancet.* **1**: 59–62.

37. Flaten, TP. (1990) Geographical associations between aluminium in drinking water and death rates with dementia (including Alzheimer's disease), Parkinson's disease and amyotrophic lateral sclerosis in Norway. *Environ. Geochem. Health.* **12**: 152–167.

38. Flaten TP. (2001) Aluminium as a risk factor in Alzheimer's disease, with emphasis on drinking water. *Brain. Res. Bull.* **55**: 187–196.

39. Neri LC, Hewitt D. (1991) Aluminium, Alzheimer's disease, and drinking water. *Lancet.* **338**: 390.

40. Rondeau V, Jacqmin-Gadda H, Commenges D *et al.* (2009) Aluminum and silica in drinking water and the risk of Alzheimer's disease or cognitive decline: Findings from 15-year follow-up of the PAQUID cohort. *Am. J. Epidemiol.* **169**: 489–496.

41. Krewski D, Yokel RA, Nieboer E *et al.* (2007) Human health risk assessment for aluminium, aluminium oxide, and aluminium hydroxide. *J. Toxicol. Environ. Health. B. Crit. Rev.* **10**, Suppl 1: 1–269.

42. Hansard. [http://hansard.millbanksystems.com/written_answers/ 1989/jul/20/lowermoor-incident-group].

43. COT (2005). Committee on Toxicity of Chemicals in Food, Consumer Products and the Environment. Subgroup Report on the Lowermoor Water Pollution Incident. Consultation Report. January 2005. [http:// cot.food.gov.uk/pdfs/lowermoorreport05.pdf]

44. Clark S. (1993) Follow-up at Camleford. *Lancet.* **341**: 488.

45. McMillan TM, Freemont AJ, Herxheimer A *et al.* (1993) Camelford water poisoning accident: Serial neuropsychological assessments and further observations on bone aluminium. *Hum. Exp. Toxicol.* **12**: 37–42.

46. McMillan TM, Dunn G, Colwill SJ. (1993) Psychological testing on schoolchildren before and after pollution of drinking water in North Cornwall. *J. Child. Psychol. Psychiatry.* **34**: 1449–1459.

47. McMillan TM. (2000) Cerebral dysfunction after water pollution incident in Camelford: Study may prolong the agony. *BMJ.* **320**: 1338.

48. Eastwood JB, Levin GE, Pazianas M *et al.* (1990) Aluminium deposition in bone after contamination of drinking water supply. *Lancet.* **336**: 462–464.

49. Owen PJ, Miles DP. (1995) A review of hospital discharge rates in a population around Camelford in North Cornwall up to the fifth anniversary of an episode of aluminium sulphate absorption. *J. Public. Health. Med.* **17**: 200–204.

50. Owen PJ, Miles DP, Draper GJ *et al.* (2002) Retrospective study of mortality after a water pollution incident at Lowermoor in north Cornwall. *BMJ* **324**: 1189.

51. Golding J, Rowland A, Greenwood R *et al.* (1991) Aluminium sulphate in water in north Cornwall and outcome of pregnancy. *BMJ* **302**: 1175–1177.

52. Altmann P, Cunningham J, Dhanesha U *et al.* (1999) Disturbance of cerebral function in people exposed to drinking water contaminated with aluminium sulphate: Retrospective study of the Camelford water incident. *BMJ* **319**: 807–811.

53. David A. (2000) Results were biased by self selection of Cases (Letters). *BMJ* **320**: 1337.

54. Esmonde TGF. (2000) Study has several methodological errors (Letters). *BMJ* **320**: 1337–1338.

55. Murray V, Goodfellow F, Bogle R. (2000) Inappropriate study, inappropriate conclusions (Letters). *BMJ* **320**: 1338.

56. McMillan TM. (2000) Study may prolong the agony (Letters). *BMJ* **320**: 1338.

57. Exley C, Esiri MM. (2006) Severe cerebral congophilic angiopathy coincident with increased brain aluminium in a resident of Camelford, Cornwall, UK. *J. Neurol. Neurosurg. Psychiatry.* **77**: 877–879.

58. Jalba DI, Cromar NJ, Pollard SJ *et al.* (2010) Safe drinking water: Critical components of effective inter-agency relationships. *Environ. Int.* **36**: 51–59.

59. Klein GL. (2005) Aluminum: New recognition of an old problem. *Curr. Opin. Pharmacol.* **5**: 637–640.

60. Klein GL, Alfrey AC, Miller NL *et al.* (1982) Aluminum loading during total parenteral nutrition. *Am. J. Clin. Nutr.* **35**: 1425–1429.

61. Klein G, Berquist W, Ament M *et al.* (1984) Hepatic aluminum accumulation in children on total parenteral nutrition. *J. Pediatr. Gastroenterol. Nutr.* **3**: 740–743.

62. Bishop NJ, Morley R, Day JP *et al.* (1997) Aluminum neurotoxicity in preterm infants receiving intravenous-feeding solutions. *N. Engl. J. Med.* **336**: 1557–1561.

63. Fewtrell MS, Bishop NJ, Edmonds CJ *et al.* (2009) Aluminum exposure from parenteral nutrition in preterm infants: Bone health at 15-year follow-up. *J. Pediatr.* **124**: 1372–1379.

64. Fewtrell MS, Edmonds CJ, Isaacs E *et al.* (2011) Aluminium exposure from parenteral nutrition in preterm infants and later health outcomes during childhood and adolescence. *Proc. Nutr. Soc.* **70**: 299–304.

65. Arnold CJ, Miller GG, Zello GA. (2003) Parenteral nutrition-associated cholestasis in neonates: The role of aluminum. *Nutr. Rev.* **61**: 306–310.

66. US. FDA. Aluminum in large and small volume parenterals used in total parenteral nutrition. Sec. 201.323 [http://www.accessdata.fda. gov/scripts/cdrh/cfdocs/cfcfr/CFRSearch.cfm?fr=201.323].

67. Graves AB, White E, Koepsell TD *et al.* (1990) The association between aluminum-containing products and Alzheimer's disease. *J. Clin. Epidemiol.* **43**: 35–44.

68. Woodson GC. (1998) An interesting case of osteomalacia due to antacid use associated with stainable bone aluminum in a patient with normal renal function. *J. Bone.* **22**: 695–698.

69. Carmichael KA, Fallon MD, Dalinka M *et al.* (1984) Osteomalacia and osteitis fibrosa in a man ingesting aluminum hydroxide antacid. *Am. J. Med.* **76**: 1137–1143.

70. Kassem M, Eriksen EF, Melsen F *et al.* (1991) Antacid-induced osteomalacia: A case report with a histomorphometric analysis. *J. Intern. Med.* **229**: 275–279.

71. Kool M, Fierens K, Lambrecht BN. Alum adjuvant: Some of the Tricks of the oldest adjuvant. *J Med. Microbiol.* (In press).

72. Tomljenovic L, Shaw CA. (2012) Mechanisms of aluminum adjuvant toxicity and autoimmunity in pediatric populations. *J. Lupus.* **21**: 223–230.

73. Lerner A. (2012) Aluminum as an adjuvant in Crohn's disease induction. *J. Lupus.* **21**: 231–238.

74. Tomljenovic L, Shaw CA. (2011) Do aluminum vaccine adjuvants contribute to the rising prevalence of autism? *J. Inorg. Biochem.* **105**: 1489–1499.

75. Aluminum therapy of silicosis. (1946) *Brit. Med. J.* June **1**: 839–840.

76. Bègin R, Massé S, Dufresne A. (1995) Further information on aluminum inhalation in silicosis. *Occup. Environ. Med.* **52**: 778–780.

77. Leung CL, Yu ITS, Chen W. (2012) Silicosis. *Lancet.* **379**: 2008–2012.

78. Penrose B. (2007) So now they have some human guinea pigs: Aluminum therapy and occupational silicosis. *Health and History* **9**: 56–79.

# CHAPTER 2

# ARSENIC

## Natural Occurrence

Arsenic is classified chemically as a metalloid having both the property of a metal and non-metal. It has an atomic number of 33 and atomic weight of 74.92. Elemental arsenic is a steel grey solid. Arsenic is the 20th most common element in the earth's crust and normally present in the environment in the inorganic form, in combination with other elements such as oxygen, sulfide and chloride to form minerals and salts. Some of the inorganic arsenic compounds are white, powder-like, tasteless and odorless. In biological systems, arsenic is found combined with carbon and hydrogen as organic arsenic. Arsenic has several oxidation states — the most common are the trivalent and pentavalent forms.[1]

Arsenic occurs naturally in the earth's crust at an average concentration of 2 mg/kg. It is mobilized by dissolution in water and emission into the atmosphere. This is accomplished naturally through weathering of minerals and igneous rock and volcanic activities. Many common arsenic compounds are soluble in water and hence can leach into river lakes and underground water. Human activities also introduce arsenic into the environment through agricultural and industrial activities and burning of coal. During smelting processes arsenic is released in the air. Ores that contain copper, lead, gold and cobalt are rich in arsenic, which is released into the atmosphere during smelting and into the environment as mining discharges. Biological systems from animals to plants and microbes also take part in the mobilization process by uptake and metabolism of arsenic.[3] The major organic arsenic compounds in the biological system are monomethylarsonous acid, monomethylarsinic acid,

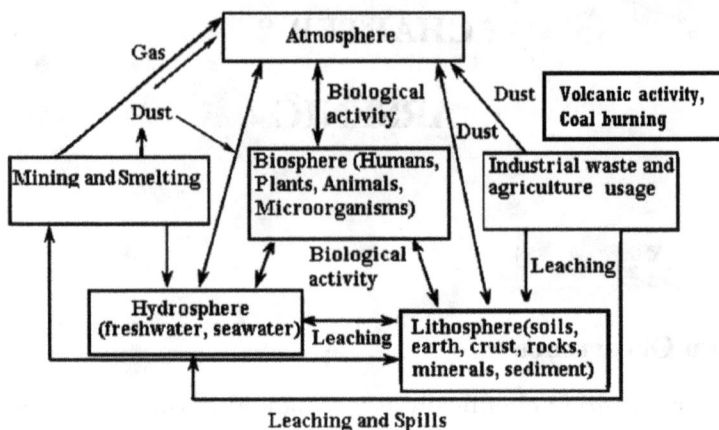

**Figure 1.**   Flow of arsenic in the global cycle.

dimethylarsinous acid, dimethylarsinic acid, trimethylarsine oxide, arsanilic acid and arsenobetane.[2]

The global arsenic cycle is a complex interactive system involving the atmosphere, hydrosphere, the earth's land mass, biosphere, volcanic activity, and human activity related to mining, smelting, industry, agriculture, coal combustion, etc. Figure 1 shows a simplified diagram of the flow of arsenic in the global ecological and biological systems.[1,4–6]

## Uses and Applications of Arsenic

Arsenic is found in pesticides, herbicides, fungicides, special paints and coatings, wood preservatives, and cotton desiccants. Organic arsenics are present in poultry and animal feeds as antimicrobials. It is used industrially in semiconductors, light emitting diodes, components of lasers and microwave circuits. It is found as alloying elements in ammunition and solders, as antifriction additives and in lead-acid battery grids.[1]

Inorganic arsenics were used as a therapeutic agent throughout the mid-19th century, primarily for the treatment of leukemia, psoriasis, and asthma; organic arsenicals were used in the treatment of protozoa diseases.[7,8] The popularity of arsenical therapeutics have waned since then; by the 1980s the only remaining organic arsenical was melarsoprol for treatment of the meningoencephalitic stage of African trypanosomiasis. Since the 1990s there has been a renewed interest in the use of arsenic trioxide for the treatment of relapsing acute promyeolcytic leukemia (See *New use for an old drug*). Realgar ($As_4S_4$), the substance employed in numerous ancient medications and ointments in both the East and the West, has recently emerged in a nanoparticle formulation as an anticancer agent.[7] Arsenic is present either as a natural ingredient or as an added component in some traditional and herbal medicines and there is concern that prolonged use may cause arsenic intoxication.[9]

## Arsenic in History

*Ancient history.* To comprehend the immensity of the impact of arsenic on human health we must begin by reaching back into its historical uses and misuses. Arsenic and substance containing arsenic were known by ancient Persians as zarnikh, meaning yellow orpiment, and by ancient Greek as arsenikon. As early as 2000 BC, arsenic trioxide, a by-product obtained from smelting copper, lead, and gold was used as a drug and a poison.[4,10,11] Orpiment ($As_2S_3$) and realgar ($As_2S_2$) which are found naturally in lead, silver and gold ore veins and sometimes near volcanic sublimation and hot spring deposits, were used by Hippocrates (460–357 BC) as escharotics and remedies for ulcers. Ancient Chinese considered Xiong Huang (arsenic sulfite) powder particularly effective cures for a variety of diseases ranging from skin infections to parasitic worms. Many of the traditional medicines dated back thousands of years in China, India and Greco–Roman time contained arsenic. It was found in all sorts of formulations and mixtures and touted as cures, tonics, ointments, perhaps even aphrodisiac.[12] Aristotle (384–322 BC) and Pliny the Elder (23–79 AD) wrote about the medical properties of arsenic. In fact, the name arsenic is derived from the Greek word "arsenikon" which means "masculine" or "potent".[11] Paracelsus, the famous 16th century physician and alchemist, was probably the first to describe in detail the preparation of metallic arsenic. He was also an advocate of its use in medicine. The use of arsenic was not only limited to the ancient world, it was a common tonic in Europe until very recently. For example, Salvarsan, a compound containing arsenic as the active ingredient, was used until 1940s to treat patients with leprosy and other skin diseases. Fowler's solution, which contained one percent potassium arsenate, was introduced by Thomas Fowler and described in the London Pharmacopeia in 1809. An apothecary label claimed that the solution is a cure for "periodic diseases" and, when applied externally, for various skin conditions including cancer. It was withdrawn from the US market in the 1950s after reports surfaced linking its prolonged use to skin changes, skin cancer and nerve damage.[13]

A review by the International Agency for Research on Cancer[14] concluded that long term medicinal use of arsenic caused a variety of cancers including those found in the skin, lungs and livers.

***New use for an old drug.*** Derek Doyle provided a concise account of the return of arsenic to clinical respectability.[65] The success of Fowler's solution in the treatment of chronic myelocytic leukemia was first reported in 1865 and then "rediscovered" in the 1931. Salvarsan, an arsenic-containing drug, was first used by Paul Erhlich in the 1900s for the treatment of syphilis and other chronic infectious diseases. In 2000, the U.S. FDA approved the use of arsenic trioxide for the treatment of chronic promyelocytic leukemia. Using sophisticated molecular biology techniques, scientists have now shown that arsenic compounds binds to and incapacitates leukemia cells. Clinical trials are now underway all over the world to determine the efficacy of these arsenic compounds.[7,15,16]

***The dose makes the poison.*** This term is definitely Paracelsus' most famous quote which means that all things when taken in excess become a poison. This is certainly true for arsenic as throughout history accidental and intentional poisoning was rife. Arsenic for domestic use as disinfectant, rat poisons, etc., was easy to obtain. Arsenic powder is readily soluble in water and the solution is colorless, odorless and tasteless and therefore difficult to detect. When given in small doses over a period of time, the symptom of arsenic poisoning is quite unremarkable and often misdiagnosed as wasting diseases or respiratory problems. For these reasons, arsenic was perhaps the most popular poison worldwide. Stories of poisoning of historical figures by arsenic are plenty. In 55 AD, the tyrannical emperor of the Roman Empire was said to have poisoned Britannicus, his rival to the throne, with arsenic.[2,13] Although the evidence was not conclusive (Nero could have easily used toxic mushroom, another popular poison of the time), it illustrated the ancient knowledge of the potency of arsenic. In middle age Europe, the Borgia family which included Pope Alexander VI and his son, Cesar, were notorious for their adroitness in using arsenic as a poison. Their victims included bishops, cardinals, and rich and famous relatives. Their use of arsenic became such a routine that they became cavalier and accidentally ingested their own arsenic-laced wine; one of them, the Pope, died. In the 16th century, Madam Signora Toffana of Palermo and Naples became the most infamous poisoner. She offered her arsenic-laden oil, along with detail instructions, to wives who wished to discretely get rid of their husbands. It was said that over 600 husbands and lovers succumbed to this *Aqua Toffana* before the madam was caught and executed. In the years since, arsenic attained such popularity that it was called *poudre de succession* or "inheritance powder". A French woman, Catherine Deshayes (1640–1680) was notorious for commercializing her expertise in arsenic poisoning. The business ended when she was convicted of numerous murders which included more than 2000 infants, and burned at the stake. As late as the 19th century, arsenic could be obtained cheaply from grocers, chemists and others "under the most frivolous pretense" and by "the lowest of the vulgar". In 1851, legislation was enacted in Britain to restrict the sale of arsenic[17] but that did not seem to be able to stop the abuse. In 1873, an English lady by the name of Mary Ann Cotton was reported to have killed more than 20 people, including her husbands and children, with arsenic. Her only benefit appeared to be the life insurance money she collected upon the death of her husbands. She was caught and hanged. Arsenic gradually lost its popularity as the poison of choice in the latter part of the 19th century. Much credit for its demise may be given to the English chemist James Marsh who, in 1836, developed a reliable test for arsenic that was so sensitive it can detect traces of arsenic in hair or nail clippings.[17] Arsenic poisoning could no longer be an invisible killer.

*A little poison is good for you?* This is an old saying the origin of which is not certain,[a] but it reflects the belief of many European of old that a small and judiciously applied dose of a poisonous substance can actually stimulates the natural defense system and invigorates the body. In more stringent term, Friedrich Nietzsche was quoted as saying: "What does not kill you makes you strong". The fabled "arsenic eaters of Styria" may be a case in point.[66] There are documentation in historical records and medical literatures indicating that this habit originated in the 12th century and carried on to the 18th century. The arsenic eaters were inhabitants of a region of what is now South–Eastern Austria. They ingested grains of arsenic trioxide (a highly toxic form of arsenic) in progressively increasing amounts because of the belief that the habit is beneficial to health and well-being — good complexion, improvement of lung function, resistance to infectious diseases, increased courage, increased sexual prowess, and more. It was believed that the practice began with feeding arsenic to horses to bolster their strength, and make their fur shiny with a healthy appearance.[12,18] Interestingly, there were no documented reports of the eaters being poisoned.

Could the arsenic eaters develop tolerance or resistance to the poison? Perhaps the native Atacameño people of Northern Chile can provide a clue. The Atacameño people have lived in the region for over 9,000 years and their source of drinking water is from rivers containing an average of 600 $\mu$g/L of arsenic (60 times the limit set by the World Health Organization). Studies showed that they suffered similar arsenic-induced skin lesions as other populations. They have increased bladder cell micronuclei which suggested that they are at increased risk of developing bladder cancer. The urinary arsenic methylation pattern of the Atacameño people was similar to other populations. In summary there were no evidence that the Atacameño people have developed resistance after hundreds of years.[19]

---

[a]The general experimental concept of a 'little poison is good for living things' has been credited to the 19th century German Scientists Hugo Schulz and Rudolf Arndt.[71]

***Simón Bolívar.*** Simón Bolívar, the Hispanic American visionary (1783–1830) and revolutionary who liberated much of the continent from Spaniard domination, was one of the most revered figures in South America. He died at a relatively young age of 47 after protracted illness which everyone assumed to be "consumption" or tuberculosis. However, there is lingering doubt that he was poisoned by political and military adversaries who accused him of being a dictator. Indeed he escaped an earlier attempt on his life thanks to the timely help of his lover Manuela Saenz. President Hugo Chavez of Venezuela, an admirer of Simón Bolívar, was an ardent advocate of the assassination by poisoning theory. He established a commission in January 2008 for exactly this purpose. This piqued the scientific curiosity of Paul Auwaerter, an infectious disease expert at Johns Hopkins School of Medicine in Baltimore, U.S.A. After carefully considering the reported symptoms, he suggested that the reported headaches, weakness, apathy, gastrointestinal complaints, coarse dark skin, and cachexia, were consistent with, although not diagnostic of, arsenicosis.[20] He further noted that arsenic-based remedies were popular during Bolivar's time, e.g. Fowler's solution was sold with claims that included cure for malaria, syphilis and many less-severe ailments. His conclusion was that either paracoccidioidomycosis or bacterial bronchiectasis complicated by chronic arsenic intoxication was the likely cause of his death rather than tuberculous consumption. In the summer of 2010, Chavez, accompanied by aids, soldiers and forensic specialist entered the National Pantheon in Caracas Venezuela and exhumed Simón Bolívar's remains. They removed several fragments of bone and teeth for study by a Presidential Commission. In December 2011, an excerpt of the Commission's finding was read on Venezuelan national television: "The hypothesis many historians had about Bolivar's death by poisoning was ruled out... the possibility that he may have consumed medicines that poisoned him unintentionally remains open".

***Napoleon Bonaparte.***[21,22,64] An even more controversial topic than that surrounding Simón Bolívar was whether Napoleon (1769–1821) was murdered by arsenic poisoning. When Napoleon died on 5 May 1821, the autopsy was conducted by his personal physician, R. Francois Carlo Antommarchi. In the autopsy report, Antommarchi wrote that the cause of death as ulcerated "cancerous" lesion in the stomach. The conclusion was never seriously challenged until 1961 when a study showed elevated arsenic concentration in one of Napoleon's hair sample which was taken at the time of his death in the small island of Saint Helena. This report triggered the interest and imagination of scientists, historians and amateurs alike, resulting in close to 30 reports on the subject and endless debates. Even Napoleon's trousers did not escape scrutiny. Alessandro Lugli et al.[21] examined trousers worn by Napoleon from 1800 to 1821 and deduced a weight gain of 23 kg between 1800 to 1820 and a weight loss of 11 kg in the last year of his life. The rapid weight loss was considered highly consistent with the course of stomach cancer. The strongest evidence against arsenic poisoning, however, was provided by nuclear physicists. The neutron activation instrument at the Italian Institute of Nuclear Physics was powerful enough to measure arsenic in a single hair. It was found that hair samples, held by museums and dated from when Napoleon was a boy to the time of his earlier exile on the Island of Elba, had high arsenic levels. Furthermore, hair samples from Napoleon's son and Empress Josephine had high arsenic levels, some as high as 100 times the concentration found in hair obtained in the present time. These results suggested that high hair arsenic appeared to be common in those days. In the final experiment, Lugli and co-workers[22] took the longest piece of hair from Napoleon, divided it into six sections, and determined their arsenic concentration. They found no evidence of a sudden increase in the arsenic content in any of the sections, as one would expect if a lethal dose had been administered during the Emperor's last six months of life. Hopefully, the poison theory is now put to rest. A surprising finding that came out of the Lugli study was that hair samples from 200 years ago contained up to 100 times more arsenic than those from today. Where did this arsenic come from? The most likely answer is that arsenic was used in everyday life at that time. It was present in face and hair powder tonics, cures, ointments, and even aphrodisiac.

The use of arsenic as a household product reached its peak during the industrial revolution when arsenic as a by-product of mining and smelting became abundant and widely available. Arsenic found new applications as rat poison, insecticide, and preservatives. It was found even in candy wrappers. The story of Paris Green, a product of the industrial revolution, deserves a special mention.

*Paris Green.* Scheele's Paris Green (copper acetoarsenite), was a popular coloring agent for paints, wall papers and fabrics in the 18th and 19th centuries. Because the main ingredient is arsenic, it has been considered the main culprit involved in many mysterious ailments and deaths in the past. Its notoriety persisted through the 1950s. Clare Booth Luce, the American Ambassador to Italy at the time, suffered a series of physical and psychological ailments which were attributed by some to arsenic poisoning. The arsenic apparently came from flaking of arsenic-laden paints on the ceiling of her residence. Another theory blamed Paris Green tinted wall papers. Molds can metabolize arsenic compounds to volatile, toxic trimethylarsine, which can be released into the room air and poisoned the inhabitants.[13]

One of the most widespread applications of Paris Green besides its use as coloring agent was as an insecticide. Our generation may not remember the intense world-wide effort in malaria control and its impacts on public health. Even fewer remembers that before the advent of DDT in the 1940s, Paris Green was the major insecticide agent used for the control of mosquito larvae and hence the spread of malaria.[23] It was even sprayed from airplanes over water-logged land. How much of the arsenic found its way into surface and ground water was anyone's guess. Because of its general accessibility at the time, Paris Green was also used for criminal purposes; a case of murder by poisoning with Paris Green was reported as recently as 1969.[24]

## Health Effects[1,2,25,26]

*Acute effects.* The earliest sign of acute arsenic poisoning is gastrointestinal: metallic or garlic-like taste in the mouth, excessive salivation, nausea, vomiting, anorexia, heartburn, abdominal pain and diarrhoea. These are followed by a broad spectrum of adverse effects that include bone marrow suppression, hematological abnormalities, capillary damages, renal failure, respiratory failure, acute hepatic toxicity, pulmonary edema, bronchial pneumonia, and dysfunction of other organs. Arsenic induced neuropathy includes encephalopathy, peripheral neuropathy, and neuritis. The acute dermal toxicity of arsenic includes dermatitis, vesiculation and melanosis. The acute cardiac cardiovascular effect of arsenic includes ventricular fibrillation, tachycardia, abnormal electrocardiograms, congestive heart failure, and hypotension. Conjunctivitis is the major ophthalmic effect of arsenic. In acute poisoning cases, the exact amount of arsenic intake was usually hard to determine; it was estimated that the minimal lethal dose is between 1–3 mg/kg body weight.

***Chronic effects.***[1,2,25,26] The chronic effect of arsenic involves all body organs. The most distinct characteristic of chronic arsenic toxicity is the skin manifestations including hyperpigmentation/depigmentation, and palmar and solar keratosis. Chronic gastrointestinal disturbance is common and is accompanied by anorexia and weight loss. Symptoms and signs of chronic neural toxicity include hearing loss, mental retardation, encephalopathy, peripheral neuropathy and neuromuscular abnormalities. Chronic hepatic effects include cirrhosis, fatty liver, and portal hypertension. Bone marrow and hematological effects include bone marrow hyperplasia, aplastic anemia, leucopenia and thrombocytopenia. There was an increase in the risk of developing lens opacity and diabetes mellitus. A common effect of chronic arsenic intoxication is cardiovascular diseases which include pericarditis, arrhythmia, electrocardiographic abnormalities, peripheral, coronary and cerebral artery diseases, carotid atherosclerosis, hypertension and microcirculation abnormalities.

***Carcinogenic effects.***[1,2,14,27] The most serious consequence of chronic arsenic poisoning is the development of cancer. There is increasing evidence, mainly from study of several regions in Taiwan (see *Blackfoot Disease in Taiwan*), that exposure to high level of arsenic in drinking water increased the risk of cancers of the skin, bladder, kidneys, prostate, lungs and liver. Markedly higher mortality rates due to cancers of the lung and bladder were observed in Chilean exposed to high arsenic in drinking water. Following exhaustive evaluation of epidemiological studies on humans and experimental studies on animals, the International Agency for Research on Cancer concluded that arsenic in drinking water is a Group I carcinogen, i.e. carcinogenic to humans.

*Mechanism of toxicity.* It is clear that arsenic has a much broader spectrum of toxic effects than many other naturally occurring chemicals and the list continues to grow.[28] A major reason is that arsenic interferes with a number of key biochemical and cellular activities. A well-known biochemical effect of arsenic is its reactivity with essential sulfhydryl groups in co-factors, enzymes, proteins and membranes. When these sulfhydryl groups or disulfide groups are bound by arsenic they cannot perform their biochemical or catalytic reactions and the normal physiological functions are disrupted. Because of their similar chemical nature, arsenic can displace phosphate in some critical biochemical reactions and thereby abort the reaction. Arsenic can interfere with redox reactions and generate reactive oxygen species. The resulting oxidative stress is thought to be detrimental to cellular functions. Recent studies also indicated that arsenic can exert its toxicity through perturbations in DNA repair and methylation, signal transduction, and cell proliferation.[2,29,30]

## Arsenic in Water

*Global contamination.* The predominant form of dissolved arsenic is trivalent and pentavalent inorganic arsenicals with the trivalent form being the more toxic. Arsenic is found in groundwater throughout the world but usually at much lower concentrations than the WHO guidance limit of 10 µg/L. There are, however, certain regions of the world that have much higher level of arsenic. A modeled map (Fig. 2) has been prepared indicating region of the globe where high levels of arsenic in groundwater may be present in either reducing or oxidizing environment.[b] The model clearly showed that many regions of the world have high probability of hav-

---

[b]Reducing condition favors release of arsenic into the aquifer. Presence of organic matters in water, for example, drives a complex series of redox reactions and promote the release of arsenic into ground waters.

(a)

(b)

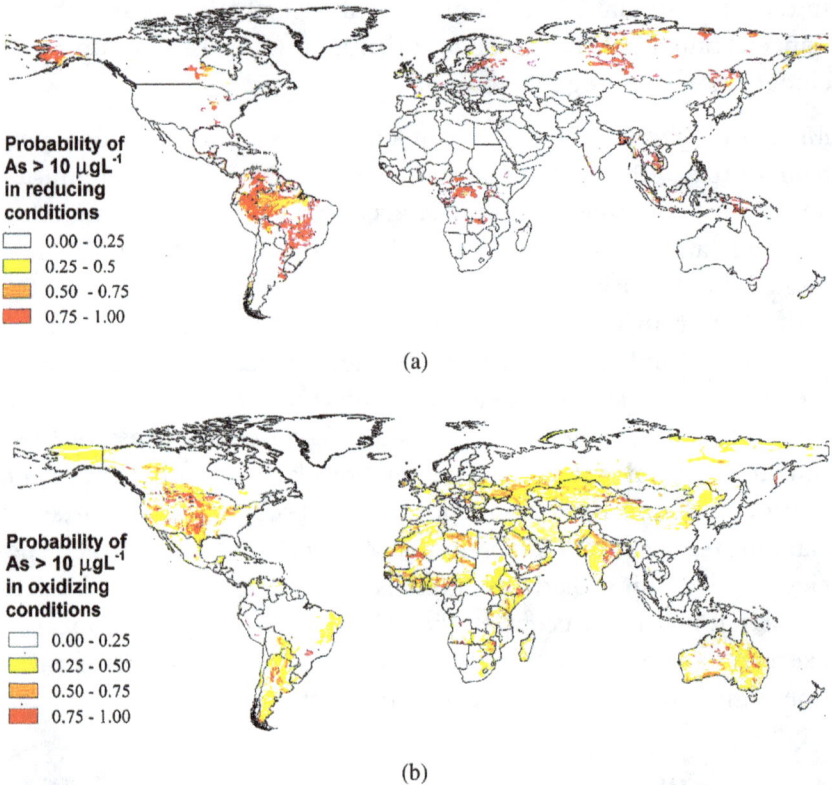

**Figure 2.** Modeled global probability of geogenic arsenic contamination.

*Source*: Amini *et al.*[34] Swiss Federal Institute of Aquatic Science and Technology, Dübendorf, Switzerland.

ing elevated level of arsenic in groundwater. Risk of human exposure to arsenic is related to the population density in the high arsenic regions. Table 1 summarized the global regions where high arsenic in groundwater is known and the estimated number of people affected. In these regions, risk of adverse health effects due of high arsenic is very real. Thirty five countries around the world have reported adverse health effects from groundwater contaminated by arsenic.[32] In the Ganga–Meghna–Brahmaputra basin alone, some 500 million inhabitants are at risk from drinking arsenic-contaminated groundwater.[33] It was estimated that about 130 million people world-wide are exposed to groundwater arsenic exceeding 50 $\mu$g/L.

In the highly contaminated regions (>500 µg/L) it was projected that 1 in 10 will die of arsenic-related diseases.[31] The complexity and immensity of the arsenic problem is illustrated by the description of three high arsenic regions of the world as follows.

***Black Foot Disease (BFD) in Taiwan.***[39–41] Arsenic poisoning is certainly not unfamiliar to the people of Taiwan in the 1950s. A well-known incidence involved 553 people who moved to a newly erected village

**Table 1.** Estimated Global Arsenic Contamination in Ground Water.[35–38]

| Country/region | Exposed population | Concentration (µg/liter) |
|---|---|---|
| Argentina | 2,000,000 | > 1 to 9,900 |
| Bangladesh | 30,000,000 | < 1 to 2,500 |
| Bolivia | 50,000 | — |
| Brazil | — | 0.4 to 350 |
| Cambodia | 100,000 | < 3,500 |
| Chile | 400,000 | 100 to 1,000 |
| Germany | — | < 10 to 150 |
| Ghana | < 100,000 | < 1 to 175 |
| Greece | 150,000 | — |
| Hungary, Romania | 400,000 | < 2 to 176 |
| Inner Mongolia, China | 100,000 to 600,000 | < 1 to 2,400 |
| Latin America (Argentina, Bolivia, Brazil, Chile, Colombia, Cuba, Ecuador, El Salvador, Guatemala, Honduras, Mexico, Nicaragua, Peru and Uruguay) | 14,000,000 | > 10 |
| Mexico | 400,000 | 8 to 620 |
| Spain | > 50,000 | < 1 to 100 |
| Taiwan | 100,000 to 200,000 | 10 to 1,820 |
| Thailand | 15,000 | 1 to > 5,000 |
| United Kingdom | — | < 1 to 80 |
| USA and Canada | — | < 1 to > 100,000 |
| Vietnam — Red River Delta | 10,000,000 | 1–3100 |
| — Mekong River Delta | 1,000,000 | ND-1351 |
| West Bengal, India | 6,000,000 | < 10 to 3,200 |
| Xinjiang, China | 50760 | < 850 |

in the south-western coast of Taiwan. Ninety percent of them contracted a curious disease: color change to the skin and hard patches on the palms and soles. They were relocated to a new village 3 km away and the symptoms subsided with no new cases appearing. The diagnosis was arsenikosis and the culprit was the contaminated well water. This is not just an isolated case. In fact, arsenic poisoning was endemic in the nearby Chia-Nan plain where inhabitants irrigate their farms and obtained their drinking water from artesian wells and springs. Many inhabitants suffered from skin conditions typical of arsenikosis. The most prominent symptom was peripheral vascular disease, mainly of the lower extremities, which resulted in discoloration, ulceration and gangrenous lesions, hence the name Black Foot Disease (BFD). The disease incidence peaked between 1956 and 1960 with prevalence rates ranging from 6.51 to 18.85 per 1,000 in different villages. The median arsenic concentration of the well water in the most affected villages ranges from 700 to 930 $\mu g/L$. After years of follow up it was found that inhabitants in this region were at higher risk of developing cancers of the liver, lungs, skin, bladder and kidneys, and the incidence increased with the concentration of arsenic and the duration of exposure. They also suffered increased mortality due to diabetes mellitus, ischemic heart disease, bronchitis, liver cirrhosis, kidney disease and cardiovascular disease. Incidence of BFD decreased dramatically in the 1970s when tap water supply was fully available in the endemic area. One mystery that remains to be solved is the uniquely high prevalence of BFD in the Chia-Nan region. In regions outside of the Chia-Nan plain which also had high arsenic in the well water, inhabitants developed the usual symptoms of arsenic poisoning but rarely the BFD. Indeed, BFD was rarely seen in other countries with high arsenic in drinking water. This difference suggested that while arsenic is the main culprit, there may be other substances also present in the groundwater in Chia-Nan that, together with arsenic, produced the characteristic BFD. In toxicology term, this is called a complex interaction. It is difficult and time consuming to tease out the primary and secondary causes when there are more than one contributory toxins, and even more difficult to elucidate the interactive effects. The artesian well water from Chia-Nan regions also contains ergot alkaloids, organic chloride and humic acid, and it has been suggested that these substances may act together with arsenic to produce BFD.[41,72] To date, there is no convincing evidence to confirm this hypothesis and the underlying mechanism of BFD remains a mystery.

***Antofagasta, Chile.***[28,42–44] Antofagasta, the second largest city in Chile with a population of over 1,000,000, and neighboring city Mejillones, were exposed to very high arsenic levels in drinking water, when their water supply was supplemented in 1958 with water from rivers that contained arsenic at concentrations close to 1,000 $\mu$g/L, which was hundred times the drinking-water limit recommended by the World Health Organization. With the installation of treatment facilities in 1970 and again in 1979, the level of arsenic ultimately dropped to 10 $\mu$g/L. Municipal and health officials at the time were advised that the new sources of water contained arsenic in excess of the standards. However, the overriding health concern at that time was diarrhea and malnutrition, and the arsenic warning was overlooked. The Chilean National Health Service had just came into existence and did not have epidemiologists in the region to initiate health surveillance. Serious health problems soon surfaced: Infant death rate started to rise soon after the peak exposure years; adults and children developed skin lesion typical of arsenicosis (leukoderma, melanoderma, and hyperkeratosis). Increased incidences of cardiovascular and respiratory diseases were reported. There were markedly higher mortality rate due to cancers of the lung and bladder. Increased childhood liver cancer mortality was also found. Almost 30 years after the high exposure period, risks of dying from cancer of the lung and bladder and from tuberculosis in Antofagasta were still markedly higher than the rest of Chile.

***Mass Arsenic Poisoning in Bangladesh.*** Diarrheal diseases caused by the consumption of pathogens contaminated surface water was a serious public health problem in Bangladesh prior to 1970s. The government of Bangladesh and international aid organizations, spearheaded by United Nations Children's Fund (UNICEF) began installing tube wells that tapped into pathogen-free aquifers as an alternative water source. The convenience and low cost of installing tube wells led millions of people to install their own private well. An estimated 40 million wells were drilled in Bangladesh.[4] It did not occur at the time that arsenic should be included in the tests to be done on water quality. It turned out the arsenic concentration was high; a 14-year study showed that 7.5%, 27.2% and 42.1% of the wells had arsenic concentration above 300, 50 and 10 $\mu$g/L, respectively.[45] Health concerns were first raised when doctors saw cases of arsenic-induced skin lesions in West Bengal, India, which is adjacent to Bangladesh. More than 1.5 million were thought to be exposed to arsenic in Bangladesh with more than 200,000 cases of poisoning.[46] Later on, physicians from Bangladesh and India also noted skin lesions in patients who came from Bangladesh.[45] Estimates vary but anywhere from 32–77 million people in Bangladesh live in regions where some wells are known to be contaminated. In neighboring West Bengal, India, another six million may be exposed to well water > 50 $\mu$g.[45-47] A World Health Organization press release[48] in 2000 warned that at least 100,000 cases of debilitating skin lesions are believed to have already occurred. It was predicted at the time that 1 in 10 who drank water with >500 $\mu$g/L arsenic will die of cancer including lung bladder and skin. Chronic diseases such as cardiovascular and pulmonary diseases were also predicted to increase with time.[46] Another estimation of cancer burden from arsenic in Bangladesh indicated at least a doubling of lifetime mortality risk from liver, bladder and lung cancers (229.6 *versus* 103.5 per 100,000 population).[49] A survey showed that poorer quality of life is associated with higher level of cumulative arsenic exposure.[50] The magnitude of the arsenic problem in Bangladesh has been described by experts as "the biggest arsenic calamity in the world"[51] and "the largest poisoning of a population in history".[46]

This human catastrophe provided stimulus for geologists to search for the origin of arsenic in the "clean" water. It was found that Bangladesh, part of India and South East Asia lie in low-lying alluvial and deltaic environment that drain the Himalayas and nearby high grounds. This water is rich in arsenic-bearing minerals originating from eroding geographical formations. The arsenic is ordinarily combined with sulfide or trapped in iron-containing minerals and is not normally found in the groundwater. However, under the right conditions, i.e. groundwater with low oxygen and high carbon, arsenic in the deposit is reduced by microbes into the soluble and more toxic AS (III) species which found their way into the aquifer. The large population and intense farming activity in Bangladesh and West Bengal provide rich organic matter which also uses available oxygen and enhances the reducing condition in groundwater.[52–54]

But all is not lost. Soon after the full impact of the catastrophe was realized, mitigation and remedial actions were initiated. These actions at the local levels included monitoring well water for arsenic and identifying tube-wells with low and high arsenic levels, installing water purification devices, and using alternative clean water supply (for example clean surface water, rain water). Recent studies suggested that tube-wells exceeding 300 m are likely to be low in arsenic and may be used for drinking. The full participation of governments at all levels as well as support from various international organizations is critical for the success of these mitigating actions. Evaluation of their impacts are still ongoing.[55,56]

Fred Pearce, author of the book *Keepers of the spring: reclaiming our water in an age of globalization*,[57] offered the following critical comment: *"Even now, as the scale of the calamity emerges, nobody is admitting culpability. Not UNICEF, which initiated the tube well program and paid for the first 900,000 wells; not the World Bank, a fellow sponsor; not the Bangladeshi government; not the foreign engineers and public health scientists who for so long did not think to test the water. The same agencies that played godmother to the catastrophe are now wringing their hands and saying it will likely take 30 years to find all the poison tubes — longer than it took to sink them all"*.

## Arsenic in Food

Arsenic is not an essential element for life.[2,58] Today, most countries prohibited the addition of arsenic in food or food supplements.

Currently the spotlight is on arsenic naturally present in foods. The predominant dietary source of arsenic is seafood, followed by rice/rice cereal, mushroom, and poultry. For fish and shellfish the arsenic is in the organic form (arsenobetaine, arsenocholine and dimethylarsinic acid) that is less harmful. Some seaweed may contain arsenic in the inorganic form which is more toxic.[1]

The U.S. has set a Minimal Risk Level of 0.3 $\mu$g/kg body weight/day based on effect of arsenic on the skin.[1] The Food and Agriculture Organization/World Health Organization (FAO/WHO)[59] has recently promulgated a new ingestion limit for inorganic arsenic, which was based on the relationship between arsenic intake and increase in incidence of lung cancer. A benchmark dose[c] for a lower limit of 0.5% increase incidence of lung cancer (BMDL$_{0.5}$) of 3 $\mu$g/kg/day (range: 2–7 $\mu$g/kg/day) was derived.

On a global scale, high arsenic content in rice in some geographic regions is a health concern. In Bangladesh, West Bengal region of India, South–Western Taiwan, and certain regions of Vietnam, Thailand and China, arsenic-contaminated groundwater is used for the cultivation of rice paddies during the dry season. As a result there is high deposition of arsenic in topsoil and high uptake of arsenic in rice grain. Total arsenic concentration in rice as high as 1.8 $\mu$g/g dry grains have been reported in China[60] and Bangladesh.[61] About 42–91% of the total arsenic in South and South–East Asian rice is the toxic inorganic species.[62] Because rice is the staple food item in these regions with a typical range of consumption of 400–650 g per day,[63] it can be calculated that the amount of arsenic ingested may exceed the WHO limit of 2–7 $\mu$g/kg body weight per day.[d] Furthermore, if the rice is cooked with water contaminated with

---

[c]A benchmark dose is a statistical dose-response analysis method using continuous dose-response data.[67] In this case the benchmark dose is 0.5% increase incidence in lung cancer.

[d]Assuming an inorganic arsenic content of 1.0 $\mu$g/g dry grain and a daily rice consumption of 500 g, the consumption of inorganic arsenic through rice consumption is 500 $\mu$g per day. An adult of 70 kg would therefore have consumed 7.1 $\mu$g inorganic arsenic/kg body weight/day.

arsenic, the total ingested arsenic will be even higher. A 2011 article indicated that arsenic in rice is the main contributor to excess cancer risk in China.[68] Concern has also been raised about higher arsenic content in baby food containing rice product.[60,69,70] More study is needed to determine the health impacts of arsenic in food.

## Practical Points

- Arsenic in water is colorless, odorless and tasteless. Boiling the water does not remove but only concentrates the arsenic in it.
- If you live in an area that is known to have high arsenic in groundwater, you should have your well water tested.
- Consult a physician if you suspect of having exposed to arsenic. Sensitive and specific tests are available that measure arsenic in your blood, urine, hair, and fingernails. These tests are helpful in determining if you have been exposed to above-average levels of arsenic.
- Arsenic must be part of the test menu for water quality.
- Limits for arsenic in drinking water varies with countries and organizations (Australia: 7 μg/L; WHO, E.U, U.S. Canada, Japan: 10 μg/L; Bangladesh, Taiwan: 50 μg/L). These are practical limits, not safety limits. Ideally, arsenic level should be as low as possible.
- Rice in certain regions of the world may have arsenic level high enough to be of concern for adults, and for infants and children in particular.
- Beware of traditional and herbal medicines that may have high levels of arsenic.[73]

## References

1. U.S. Department of Health and Human Services. Public Health Service. Agency for Toxic Substances and Disease Registry. (2007) Toxicological Profile for Aluminum. Atlanta, Georgia.
2. Hughes MF, Beck BD, Chen Y *et al.* (2011) Arsenic exposure and toxicology: A historical perspective. *Toxicol. Sci.* **123**: 305–332.

3. Duker AA, Carranza EJ, Hale M. (2005) Arsenic geochemistry and health. *Environ. Int.* **31**(5): 631–641.

4. Mudhoo A, Sharma SK, Garg VK *et al.* (2011) Arsenic: An Overview of Applications, Health, and Environmental Concerns and Removal Processes. *Crit. Rev. Environ. Sci. Technol.* **41**: 435–519.

5. Langdon CJ, Piearce TG, Meharg AA *et al.* (2003) Interactions between earthworms and arsenic in the soil environment: A review. *Environ. Pollut.* **124**: 361–373.

6. Shih MC. (2005) An overview of arsenic removal by pressure-driven membrane processes. *Desalination* **172**: 85–97.

7. Wu J, Shao Y, Liu J *et al.* (2011) The medicinal use of realgar (As4S4) and its recent development as an anticancer agent. *J. Ethnopharmacol.* **135**: 595–602.

8. Nicolis I, Curis E, Deschamps P *et al.* (2009) Arsenite medicinal use, metabolism, pharmacokinetics and monitoring in human hair. *Biochimie.* **91**: 1260–1267.

9. Martena MJ, Van Der Wielen JC, Rietjens IM *et al.* (2010) Monitoring of mercury, arsenic, and lead in traditional Asian herbal preparations on the Dutch market and estimation of associated risks. *Food. Addit. Contam. Part. A. Chem. Anal. Control. Expo. Risk. Assess.* **27**: 190–205.

10. Nriagu JO. (2002) Arsenic poisoning through the ages. In: Frankenberger WT (ed.), *Environmental Chemistry of Arsenic.* Marce, Dekker, NY.

11. Jolliffe DM. (1993) A history of the use of arsenicals in man. *J. R. Soc. Med.* **86**: 287–289.

12. Cullen WR. (2008) *Is Arsenic an Aphrodisiac? The Sociochemistry of an Element.* Royal Society of Chemistry, Cambridge, U.K.

13. Gorby MS. (1988) Arsenic poisoning (Clinical Conference). *West. J. Med.* **149**: 308–315.

14. IARC (2004) *Monographs on the Evaluation of Carcinogenic Risks to Humans. Some Drinking-water Disinfectants and Contaminants, Including Arsenic. International Agency for Research on Cancer,* Vol. 84. IARC, Lyon, France.

15. Wu J, Shao Y, Liu J *et al.* (2011) The medicinal use of realgar ($As_4S_4$) and its recent development as an anticancer agent. *J. Ethnopharmacol.* **135**: 595–602.

16. Lengfelder E, Hofmann WK, Nowak D. (2012) Impact of arsenic trioxide in the treatment of acute promyelocytic leukemia. *Leukemia* **26**: 433–42.

17. Bartrip P. (1992) A "pennurth of arsenic for rat poison": The Arsenic Act, 1851 and the prevention of secret poisoning. *Med. Hist.* **36**: 53–69.

18. Przygoda G, Feldmann J, Cullen WR. (2001) The arsenic eaters of Styria: A different picture of people who were chronically exposed to arsenic. *Appl. Organomet. Chem.* **15**: 457–462.

19. Smith AH, Arroyo AP, Mazumder DN *et al.* (2000) Arsenic-induced skin lesions among Atacameño people in Northern Chile despite good nutrition and centuries of exposure. *Environ. Health. Perspect.* **108**: 617–20.

20. Auwaerter PG, Dove J, Mackowiak PA. (2011) Simon Bolivar's medical labyrinth: An infectious diseases conundrum. *Clin. Infect. Dis.* **52**: 78–85.

21. Lugli A, Lugli AK, Horcic M. (2005) Napoleon's autopsy: New perspective. *Human. Pathol.* **36**: 320–324.

22. Lugli A, Clemenza M, Corso PE *et al.* (2011) The medical mystery of Napoleon Bonaparte: An interdisciplinary exposé. *Adv. Anat. Pathol.* **18**: 152–158.

23. McWilliams JE. (2008) "The horizon opened up very greatly": Leland O. Howard and the transition to chemical insecticides in the United States, 1894–1927. *Agric. Hist.* **82**(4): 468–495.

24. Kamien M. (2010) A diagnostic challenge in peripheral neuropathy. *Clin. Med.* **10**: 245.

25. Ratnaike RN. (2003) Acute and chronic arsenic toxicity. *Postgrad. Med. J.* **79**: 391–396.

26. Chen C-J. (2010) Health hazards of arsenic in drinking water. In: Jean, Bundschuh and Bhattacharya (eds.), *Arsenic in Geosphere and Human Disease*, pp. 251–253. Taylor & Francis, London.

27. Schuhmacher-Wolz U, Dieter HH, Klein D *et al.* (2009) Oral exposure to inorganic arsenic: Evaluation of its carcinogenic and non-carcinogenic effects. *Crit. Rev. Toxicol.* **39**: 271–298.

28. Smith AH, Steinmaus CM. (2010) Arsenic in drinking water. *BMJ.* **342**: 2248.

29. Schuhmacher-Wolz U, Dieter HH, Klein D *et al.* (2009) Oral exposure to inorganic arsenic: Evaluation of its carcinogenic and non-carcinogenic effects. *Crit. Rev. Toxicol.* **39**: 271–298.

30. Druwe IL, Vaillancourt RR. (2010) Influence of arsenate and arsenite on signal transduction pathways: An update. *Arch. Toxicol.* **84**: 585–596.

31. Van Halem D, Bakker SA, Amy GL *et al.* (2009) Arsenic in drinking water: A world wide quality concern for water supply companies. *Drink. Water. Eng. Sci.* **2**: 29–34.

32. Mukherjee A, Sengupta MK. (2006) Arsenic contamination in groundwater: A global perspective with special emphasis on the Asian scenario: Special issue on arsenic. *J. Health. Popul. Nutr.* **24**: 142–163.

33. Chakraborti D. (2004) Groundwater arsenic contamination and its health effects in the Ganga-Meghna-Brahmaputra plain. *J. Environ. Monitor.* **6**: 74N–83N.

34. Amini M, Abbaspour KC, Berg M *et al.* (2008) Statistical modeling of global geogenic arsenic contamination in groundwater. *Environ. Sci. Technol.* **42**: 3669–3675.

35. Nordstrom DK. (2002) Public health: Worldwide occurrences of arsenic in ground water. *Science.* **296**: 2143–2145.

36. Ng J, Liu FF, Wang JP *et al.* (2010) Biomarkers for evaluation of population health status 16 years after the intervention of arsenic-contaminated groundwater in Xinjiang, China. In Jean, Bundschuh and Bhattacharya (eds.), *Arsenic in Geosphere and Human Diseases,* pp. 288–290. Taylor & Francis, London.

37. Kim KW, Chanpiwat P, Hanh HT *et al.* (2011) Arsenic geochemistry of groundwater in Southeast Asia. *Front. Med.* **5**: 420–433.

38. Bundschuh J, Litter MI, Parvez F *et al.* (2011) One century of arsenic exposure in Latin America: A review of history and occurrence from 14 countries. *Sci. Total. Environ.* In press.

39. Tseng WP. (1977) Effects and dose-response relationships of skin cancer and blackfoot disease with arsenic. *Environ. Health. Perspect.* **119**: 109–119.

40. Tseng CH. (2005) Blackfoot disease and arsenic: A never-ending story. *J. Environ. Sci. Health. C. Environ. Carcinog. Ecotoxicol. Rev.* **23**: 55–74.

41. Gau RJ, Yang HL, Suen JL *et al.* (2001) Induction of oxidative stress by humic acid through increasing intracellular iron: A possible mechanism leading to atherothrombotic vascular disorder in blackfoot disease. *Biochem. Biophys. Res. Commun.* **283**: 743–749.

42. Ferreccio C, Sancha AM. (2006) Arsenic exposure and its impact on health in Chile. *J. Health. Popul. Nutr.* **24**(2): 164–175.

43. Liaw J, Marshall G, Yuan Y *et al.* (2008) Increased childhood liver cancer mortality and arsenic in drinking water in northern Chile. *Cancer. Epidemiol. Biomarkers. Prev.* **17**: 1982–1987.

44. Smith AH, Marshall G, Yuan Y *et al.* (2011) Evidence from Chile that arsenic in drinking water may increase mortality from pulmonary tuberculosis. *Am. J. Epidemiol.* **173**: 414–420.

45. Chakraborti D, Rahman MM, Das B *et al.* (2010) Status of groundwater arsenic contamination in Bangladesh: A 14-year study report. *Water Research* **44**: 5789–5802.

46. Smith AH, Lingas EO, Rahman M. (2000) Contamination of drinking-water by arsenic in Bangladesh: A public health emergency. *Bull. World. Health. Organ.* **78**: 1093–1103.

47. Smedly PL, Kinniburgh DG. Chapter 1. Source and behaviour of arsenic in natural waters. In *United Nations Synthesis Report on Arsenic in Drinking Water.* [http://www.who.int/water_sanitation_health/dwq/arsenic3/en/index.html].

48. World Health Organization (2000) Researchers warn of impending disaster from mass arsenic poisoning. [http://www.who.int/inf-pr-2000/en/pr2000-55.html]

49. Chen Y, Ahsan H. (2004) Cancer burden from arsenic in drinking water in Bangladesh. *Am. J. Public. Health.* **94**: 741–744.

50. Laskar MS, Rahaman MM, Akhter A *et al.* (2010) Quality of life of arsenicosis patients in an arsenic-affected rural area in Bangladesh. *Int. Arch. Occup. Environ. Health.* **65**: 70–76.

51. Dhar RK, Biswas BK, Samanta G *et al.* (1998) Groundwater arsenic contamination and sufferings of people in Bangladesh may be the biggest arsenic calamity in the world. International Conference on Arsenic Pollution of Groundwater in Bangladesh: Cause, effects and remedies, Dhaka, Bangladesh, February 8–12: 86–87.

52. Harvey CF, Swartz CH, Badruzzaman AB *et al.* (2002) Arsenic mobility and groundwater extraction in Bangladesh. *Science* **298**: 1602–1606.

53. Harvey CF. (2008) Environmental science: Poisoned waters traced to source. *Nature* **454**: 415–416.

54. Fendorf S, Michael HA, van Geen A. (2010) Spatial and temporal variations of groundwater arsenic in South and Southeast Asia. *Science* **328**: 1123–1127.

55. Rammelt CF, Boes J. (2004) Arsenic mitigation and social mobilization in Bangladesh. *Int. J. Sustain. High. Educ.* **5**: 308–319.

56. Gardner R, Hamadani J, Grandér M *et al.* (2011) Persistent exposure to arsenic via drinking water in rural Bangladesh despite major mitigation efforts. *Am. J. Public. Health.* **101**: S333–S338.

57. Pearce F. (2004) *Keepers Of The Spring: Reclaiming Our Water in an Age of Globalization.* Island Press, Washington, USA.

58. National Research Council. (1999) *Arsenic in Drinking Water.* National Academy Press, Washington, D.C.

59. FAO/WHO (2010) Joint FAO/WHO Expert Committee on Food Additives. Seventy-second Meeting. Summary and Conclusions. [http://www.who.int/foodsafety/chem/summary72_rev.pdf].

60. Sun GX, Williams PN, Carey AM *et al.* (2008) Inorganic arsenic in rice bran and its products are an order of magnitude higher than in bulk grain. *Environ. Sci. Technol.* **42**: 7542–7546.

61. Meharg AA, Rahman MM. (2003) Arsenic contamination of Bangladesh paddy field soils: Implications for rice contribution to arsenic consumption. *Environ. Sci. Technol.* **37**: 229–234.

62. Rahman MA, Hasegawa H. (2011) High levels of inorganic arsenic in rice in areas where arsenic-contaminated water is used for irrigation and cooking. *Sci. Total. Environ.* **409**: 4645–4655.

63. Duxbury JM, Mayer AB, Lauren JG *et al.* (2003) Food chain aspects of arsenic contamination in Bangladesh: Effects on quality and productivity of rice. *J. Environ. Sci. Health. A. Tox. Hazard. Subst. Environ. Eng.* **38**: 61–69.

64. Forschuvud S, Smith H, Wassen A. (1961) Arsenic content of Napoleon I's hair probably taken immediately after his death. *Nature.* **192**: 103–105.

65. Doyle D. (2009) Notoriety to respectability: A short history of arsenic prior to its present day use in haematology. *Brit. J. Haematol.* **145**: 309–317.

66. MacLaglan C. (1864) On the arsenic-eaters of Styria. *Boston. Med. Surb. J.* **71**: 198–205.

67. Crump K. (2002) Critical issues in benchmark calculations from continuous data. *Crit. Rev. Toxicol.* **32**: 133–153.

68. Li G, Sun GX, Williams PN *et al.* (2011) Inorganic arsenic in Chinese food and its cancer risk. *Environ. Int.* **37**: 1219–1225.

69. Burló F, Ramírez-Gandolfo A, Signes-Pastor AJ *et al.* (2011) Arsenic contents in Spanish infant rice, pureed infant foods, and rice. *J. Food. Sci.* **77**: T15–T19.

70. Jackson BP, Taylor VF, Karagas MR *et al.* (2012) Arsenic, organic foods, and brown rice syrup. *Environ. Health. Perspect.* In press.

71. Calabrese EJ, Baldwin L. (2000) Chemical hormesis: Its historical foundations as a biological hypothesis. *Human and Experimental Toxicology* **19**: 2–31.

72. Yen CC, Lu FJ, Huang CF, *et al.* (2007) The diabetogenic effects of the combination of humic acid and arsenic: *in vitro* and *in vivo* studies. *Toxicol. Lett.* **172**: 91–105.

73. Wong SS, Tan KC, Goh CL. (1998) Cutaneous manifestations of chronic arsenicism: review of seventeen cases. *J. Am. Acad. Dermatol.* **38** (2 Pt 1): 179–85.

# CHAPTER 3

# COPPER

## Occurrence and Use

Pure copper is a reddish-brown and soft metal that is easily molded and shaped. It has an atomic number of 29 and an atomic weight of 63.5. It belongs to group 11 of the Periodic Table along with other so called coinage metals such as silver and gold. It is ubiquitously present in the environment in the metallic form and as mineral salts and organic and inorganic compounds. Many copper compounds can be recognized by their blue-green colors. On average, the earth's crust contains 50 part per million (ppm) of copper. Cupric ion (Cu(II)) is the most important oxidation state of copper and is the state generally encountered in water.[1] The other oxidation states of copper are Cu (I), Cu (III) and Cu (O). Generally speaking the soluble form of copper is more toxic than the insoluble or particulate forms.[1,2]

Copper is an essential element for life. Our body cannot function without it. Many of the enzymes whose catalytic actions are necessary for life need copper to work. Copper is so important that our body has several proteins that deal with transportation, transfer and storage of copper. Copper is toxic only when there is a huge excess or, in rare cases, when genetic defects make the body incapable of handling copper.

Products containing copper, e.g. electrical wiring, cookware, and plumbing, are commonly found in households. Most utensils and appliances containing copper are coated to prevent the copper from being oxidized. Copper sheets are popular building material for roofing and architectural and decorative purposes. The elegant green color on the bronze roof of buildings is actually copper carbonate that is formed on the surface of the sheet as copper is gradually oxidized by the moist air. The industrial uses of copper include electrical and electronic products, heat exchangers and valves and fittings. Copper is present in some wood preservatives. Copper salts are used as algaecides, fungicides and molluscides. It is well known to researchers working with cell cultures that a few drop of copper sulfate solution is the best and cheapest antimicrobial agent. Copper may be added in trace amount in some food and health supplements. The primary component of some intra-uterine devices is the copper coil. The contraceptive action is due to a combination of the induced localized inflammatory condition and the effect of the copper ions on sperm viability and mobility.[3]

## History

Copper is intimately involved in human civilization. The period between approximately 5,000–3,500 BC when humans started to use copper was called the Chalcolithic age. The Bronze Age (3,500–600 BC) began when other metals, chiefly tin, were added to copper to produce bronze, a much harder alloy. It was known that around 5000 BC, people living in what is now Turkey mastered the technique of smelting and refining copper to make ornaments, statues and utensils. In Northern Negev and Palestine, archaeologists have uncovered sites and artifacts indicating there were cities during the Chalcolithic period that engaged in smelting copper and even some copper alloys. More recent evidence showed that before 5000 BC, ancient settlers in what is now Serbia had mined ore from nearby sources, melted and extracted the copper and fashioned tools and figurines from it.[4] The famous Ötzi man, whose frozen body was discovered in the Alps in 1991 and dated to around 3,300 BC, carried an axe with a blade of rare copper which marked him as a man of status.[5] His hair was reported to have elevated levels of copper and arsenic which suggest that he might have been a copper smelter. Copper was so important a resource that ancient cities were named after it. For example, Cyprus, from the Latin cyprium for copper, was an important copper mining and smelting center since the early Bronze period. The city of Tung Ling (copper capital in Chinese) in Eastern China was named after its smelting industry with continuous recorded history dating from before 1,000 BC to 300 AD.

The oldest continuously mined copper ore in the ancient world is probably Wadi Faynan in Southern Jordan. The cite was inhabited from the Chalcolithic age until the Roman period. During that time it was extensively mined using prisoners, slaves and, later on, captured Christians. What was once a prosperous Roman industrial centre is now a ruin heavily contaminated with copper and lead. How bad was the copper pollution in ancient time? Archaeologists showed that skeletons excavated from Wadi Fanynan had average copper concentration of 52.6 ppm, more than 80 times higher than the level found in bones from a modern industrial area.[6,7] Grattan quoted Bishop Eusebius of Caesarea who described the work site as a place "*where even a condemned murderer may live only a few days*". It is not difficult to imagine that the miners, slaves and guards at the time became progressively weakened by inhaling particulates contaminated with copper, and by consuming food enriched with copper. On top of this, the land itself became barren as crops failed to thrive in the contaminated soil.[8] Even though the mining activity ceased 1,500 years ago, the contamination situation did not disappear with time. Today in the vicinity of Wadi Faynan where Bedouins sometimes pitched their tents, copper concentration in the soil reached 2,849 ppm. Plants and animals consumed by them also contained high levels of copper. People are still suffering from the consequences of copper pollution generated thousands of years ago.

## Health Effects

Copper is an essential element and our body cannot function properly without it. It is an essential part of many enzymes involved in hemoglobin formation, drug/xenobiotic metabolism, carbohydrate metabolism, catecholamine biosynthesis, cross-linking of collagen, elastin, and hair keratin, and the antioxidant defense mechanisms.

We can appreciate the importance of copper by its distribution in our vital organs. The highest concentrations of copper in tissues are found in the brain and the liver. Liver is the major storage and mobilization depot for copper. On the other hand, the high concentration of copper in the brain points to its critical role in neurotransmitter metabolism.[9] The importance of copper in the nervous system can be seen when copper is experimentally depleted. When animals were treated with cuprizone, a copper chelating (binding) agent, tissue copper is depleted and they developed neurodegenerative diseases. Further examination showed that the diseases are due to central nervous system demyelination. Cuprizone is now used to generate animal model for the study of multiple sclerosis and schizophrenia.[10,11]

*Diseases associated with copper.* We can also learn about the effect of copper on our body by looking at some inherited diseases related to defect in gene products involved in copper metabolism.

Wilson disease is an inherited autosomal disease caused by mutation in gene encoding a copper transporting protein ATP7B. The prevalence of this disease is estimated to be 1 in every 30,000 live births. The defective protein causes deficiency in incorporation of copper into enzymes and dysfunction in the secretion of copper into the bile. The resultant accumulation of copper causes severe liver damages. Neurological and psychiatric symptoms may also be present patients with Wilson disease.[12]

Menkes disease is an inherited neurodegenerative disorder present in approximately 1 in every 10,000 births, with epilepsy and brain degeneration as the major clinical and pathological presentations.[13] The disease originates from a mutation in the gene coding for a key copper transportation protein, ATP7A. Impaired copper transportation across specific tissues causes deficiency in copper-containing enzymes and widespread disruption of cellular functions, principally neurotransmitter and energy metabolism. ATP7A is involved in transportation of copper across specific tissues including the brain, hence the neurological presentation.

There are two other diseases caused by chronic copper toxicity whose etiology is not yet clearly elucidated because the participation of a genetic component is not certain. One is the Indian childhood cirrhosis which occurred most often in infants and young children in the Indian subcontinent and is characterized by high levels of copper in tissues and liver cirrhosis. The external cause is the consumption of animal milk stored or heated in copper or copper-alloy utensils. The condition dramatically diminished after the population was instructed to avoid using copper vessels to store and heat milk.[14] The second is idiopathic chronic toxicosis,[14,15] an extremely rare disease occurring in 1 in 1,000,000 births in some regions of Germany and Austria. There is very little known about the disease except that there may be a genetic element of susceptibility towards copper. Similar to the Indian childhood cirrhosis, the condition dramatically diminished after the population was advised to avoid using copper vessels to store and heat milk.

What do these diseases tell us? Both Wilson and Menkes diseases are rare but fatal and both are caused by defective copper transportation, utilization and disposal, thus illustrating the important role of copper play in biological functions. Small changes in the amount of copper entering our body through exposure to the environment and diets are probably not harmful. However, for people who are genetically predisposed to defects in copper transportations and metabolism, elevated amount of copper intake is harmful and even fatal.

**Copper and Alzheimer's disease?** Alzheimer's disease (AD) is a progressive degenerative disease of the brain. An early sign of the disease is the gradual loss of cognitive function. The disease is caused by the abnormal deposition in the brain of amyloidal plaques constituting primarily β amyloid, a copper/zinc metalloprotein. It was estimated that about 33.9 million people worldwide have AD, and this prevalence is expected to triple over the next 40 years.[16]

The original observation that led to the putative association of copper with AD originated with the research of Sparks and colleagues who produced an AD-like condition in New Zealand's white rabbits by feeding them with a diet high in cholesterol.[17] After they moved the laboratory to a new location they were surprised that the result could not be reproduced when the experiment was repeated. After much sleuthing they narrowed down the difference in experimental procedures to the water given to the animals — tap water containing appreciable amount of copper was used in the first laboratory and pure distilled water in the new laboratory. Further study showed that as little as 0.12 ppm copper given to cholesterol-fed rabbits induced β-amyloid accumulation and retarded specific learning ability.[18] This finding, if confirmed, has alarming health implications because the present guideline limits for copper in drinking water set by the U.S. Environmental Protection Agency[19] is 1.3 ppm, and the provisional limit set by the World Health Organization[20] is 2 ppm, many times higher than the 0.12 ppm that was found to induce AD-like condition in cholesterol-fed rabbits. Subsequent studies carried out by other groups focused on finding out the mechanisms behind this copper effect. Some investigators proposed that copper is involved in the pathogenesis of AD by influencing the aggregation of the β amyloid protein and hence its deposition in the brain.[21] Others suggested that oxidative stress associated with the formation of reactive oxygen species by copper-amyloid complexes contributed to the neurodegeneration processes in AD patients.[22]

A team of epidemiologists led by Rosanne Squitti [23–25] reported that free copper in AD patients was higher compared to age-matched controls and that AD patients with higher free copper also scored lower in a test measuring cognitive functions. They extended the observation by reporting that patients with mild cognitive impairment also had significantly higher free copper than healthy controls. It should be noted that these epidemiology studies were conducted on a very small group of patients and therefore need to be verified with studies involving larger group of patients and controls. As Squitti and her colleague pointed out, the role of copper in AD is still controversial. Indeed in their overview of all existing studies they have found that free copper in serum was significantly associated with AD but copper in plasma and cerebrospinal fluid were not[26] — a puzzling discrepancy that needs to be explained. In a review on copper toxicity, Brewer[27] pointed out that all of the molecules known to be involved with the pathology of AD and certain risk factors for AD are binders of copper. These included all the transporters of copper, various enzymes and, of course, β amyloid. In short, the copper theory is attractive and there are many champions. So far, epidemiological association of copper with AD came from only a limited number of research centers. It should be noted that there are many non-copper risk factors e.g. aluminum (Chapter 1) known to be associated with AD.[40] It will take more well-controlled, multidisciplinary studies to verify or refute the copper claim. In the meantime there are already debates. On the approach to investigate the putative effects of copper on AD. Copper overload, copper deficiency or imbalance copper homeostasis have all had their proponents.[41,42] At this point in time we can only wait patiently for the truth to emerge.

***Acute Health Effects.*** Medical literature is replete with case reports of copper poisoning; four representative cases are described below:

> A group of school children drank a lime-based beverage which was stored overnight in an urn. They suffered nausea, vomiting, abdominal pain and diarrhea. The beverage later was found to be contaminated with copper leaching from the urn. The situation was aggravated by the high acidity of the beverage which promotes the solubilization of copper over a long period of storage time.[28]
>
> Twelve people who worked on a small industrial estate suffered abdominal pain and vomiting symptoms. Further investigation showed that copper pipes in the distribution system showed evidence of internal corrosion and blue crystals were noted precipitating out of a routine water sample. Analysis of the sample gave a copper concentration of 45 ppm.[29]
>
> Eleven of 29 patients hospitalized with copper sulfate poisoning developed acute kidney failure. They were estimated to have ingested from 1 to 50 g copper sulfate. Intravascular hemolysis (lysis of red blood cells inside the circulation) and the ensuing clogging of the kidneys appeared to be the chief factor responsible for the condition. Five of the 11 patients did not survive.[30]

In Wisconsin, USA, 37 persons from two private homes became ill due to copper poisoning. The drinking water was found to have high concentration of copper due to leaching from the newly installed copper plumbing.[31]

These cases illustrated the common conditions leading to acute copper poisoning: beverages and drinking waters contaminated with copper and accidental or intentional ingestion of copper sulfate. Copper sulfate is widely used as fungicides, algaecides, and molluscides. It was used in the past as an emetic and still can be found in school chemistry sets and hobby kits such as those for crystal growing. Health effects ranged from nausea, vomiting, stomach pain, diarrhea to intravascular hemolysis and liver and kidney changes. In extreme cases, death may occur. In the Indian subcontinent and probably many other countries poisoning by copper sulfate, or blue vitriol as it is called, is still quite common with a mortality of 22.9%.[32]

***Long Term Health Effects.*** There is only a limited number of reports on the effect of chronic exposure to copper on health effects. In one case, a family reported recurrent episodes of gastrointestinal illness involving vomiting and abdominal pain after drinking water drawn from the kitchen faucet. One early-morning water sample taken from the family household contained a copper level of 7.8 ppm, which is above the US standard for drinking water (1.3 ppm). For over a year the household's water was drawn from a copper main and it was suspected that copper levels increased when the water remained stagnant in the main. All symptoms resolved when they stopped drinking water from home.[33] Gastrointestinal symptoms (nausea, vomiting, abdominal pain, and diarrhea) appeared to be the main effects following long term consumption of drinking water with high concentration of copper.[34,35]

## Practical Points

***Water.*** The most probable route of long-term exposure to high level of copper is through drinking water. The guideline for copper in drinking water of 1.3 ppm (US EPA) or 2 ppm (WHO) was established to provide an adequate margin of safety in population with normal copper homeostasis. It is prudent for patients with deficiency in copper homeostasis, such as Wilson disease, Menkes disease, Indian childhood cirrhosis and idiopathic childhood toxicosis, to avoid food and drink with high copper concentration.

One should avoid storing acidic water and beverage in copper and copper-alloy containers. Corrosion of copper pipes and storage tank are common sources of copper contamination. In area where water is acidic or rich in carbonate, the corrosion and solubilization problems are even more serious.

Old water pipes, faucets and containers should be regularly checked for corrosion. A faucet that has not been used for more than 6 hours should be flushed with running water for at least 15 seconds. Water that has been in storage for a long time may have higher copper levels. It is therefore a safe practice to fully flush the system and use running water only. Avoid using hot water for drinking and cooking purpose as it tends to dissolve more copper than cold water.

The thresholds for tasting copper in drinking water varied widely from person to person,[36,37] and should not be relied on for sensing copper contamination. Furthermore there is a danger of adaptation when one become desensitized to the presence of copper taste after prolonged use. When copper contamination is suspected water specimens should be sent to a certified laboratory for testing.

Boiling or bleaching with chlorine will not remove copper from water. The best way to remove copper from water is by treatment methods such as reverse osmosis, distillation, or ion exchange.

***Total Oral Intake.*** The second most common cause of high copper exposure is from food consumption. The daily diet of an adult usually provides around 2–5 mg of copper.[43] However some food items contain very high level of copper. For example, liver contain in excess of 100 ppm of copper and oysters contain anywhere from 25 to 600 ppm copper. If shellfish is harvested from area polluted with metals, the copper concentration can be even higher. Regular consumption of copper-rich shellfish may contribute as much as 150 mg of copper to a daily diet.[2]

The U.S. Food and Nutritional Board[38] estimated the recommended dietary intake and the upper safety level for copper to be 0.9 mg/day and 10 mg/day, respectively. A new dose-response model that takes into account the adverse effects of copper deficiency and the toxicity of excessive copper resulted in an optimal intake level of 2.6 mg Cu/d.[39] These estimated values are used by health and regulatory agencies to ensure that copper in food and water we consumed are within a safe limit.

# References

1. Cotton FA, Wilkinson G. (1980) *Copper. Advanced Inorganic Chemistry*, pp. 798–821. John Wiley and Sons, New York, NY.
2. Toxicological Profile for Copper. (2004) US Department of Health and Human Services. Public Health Service. Agency for Toxic Substances and Disease Registry. Atlanta, Georgia 30333.
3. Stanford JB, Mikolajczyk RT. (2002) Mechanisms of action of intrauterine devices: Update and estimation of postfertilization effects. *Am. J. Obstet. Gynecol.* **187**: 1699–1708.
4. Radivojević M, Rehren T, Pernicka E *et al.* (2010) On the origins of extractive metallurgy: New evidence from Europe. *J. Archaeol. Sci.* **37**: 2775–2787.
5. National Geographic. Unfrozen. November 21 issue. pp. 118–133.
6. Grattan J, Huxley S, Abu Karaki L *et al.* (2002) 'Death ... more desirable than life'? The human skeletal record and toxicological implications of ancient copper mining and smelting in Wadi Faynan, southwestern Jordan. *Toxicol. Ind. Health.* **18**: 297–307.
7. Grattan JP, Huxley SI, Payatt FB. (2003) Modern Bedouin exposure to copper contamination: An imperial legacy? *Ecotox. Environ. Safe.* **55**: 108–115.
8. Pyatt FB, Barker GW, Birch P *et al.* (1999) King Solomon's Miners — starvation and bioaccumulation? An environmental archaeological investigation in Southern Jordan. *Ecotox. Environ. Safe.* **43**: 305–308.

9. Uauy R, Maass A, Araya M. (2008) Estimating risk from copper excess in human populations. *Am. J. Clin. Nutr.* **88**: 867S–871S.

10. Herring NR, Konradi C. (2011) Myelin, copper, and the cuprizone model of schizophrenia. *Front. Biosci.* (*Schol Ed*). **3**: 23–40.

11. Pott F, Gingele S, Clarner T *et al.* (2009) Cuprizone effect on myelination, astrogliosis and microglia attraction in the mouse basal ganglia. *Brain. Res.* **1305**: 137–149.

12. Weiss KH, Stremmel W. (2010) Evolving perspectives in Wilson disease: Diagnosis, treatment and monitoring. *Curr. Gastroenterol. Rep.* **14**: 1–7.

13. Prasad AN, Levin S, Rupar CA *et al.* (2011) Menkes disease and infantile epilepsy. *Brain. Dev.* **10**: 866–876.

14. de Romaña DL, Olivares M, Uauy R *et al.* (2011) Risks and benefits of copper in light of new insights of copper homeostasis. *J. Trace. Elem. Med. Biol.* **25**: 3–13.

15. Müller T, Schäfer H, Rodeck B *et al.* (1999) Familial clustering of infantile cirrhosis in Northern Germany: A clue to the etiology of idiopathic copper toxicosis. *J. Pediatr.* **135**: 189–196.

16. Barnes DE, Yaffe K. (2011) The projected effect of risk factor reduction on Alzheimer's disease prevalence. *Lancet. Neurol.* **10**: 819–828.

17. Sparks DL, Lochhead J, Horstman D *et al.* (2002) Water quality has a pronounced effect on cholesterol-induced accumulation of Alzheimer amyloid beta (Abeta) in rabbit brain. *J. Alzheimers. Dis.* **4**: 523–529.

18. Sparks DL, Schreurs DB. (2003) Trace amounts of copper in water induce β-amyloid plaques and learning deficits in a rabbit model of Alzheimer's disease. *Proc. Natl. Acad. Sci. U S A.* **100**: 11065–11069.

19. US Environmental Protection Agency. (1991) 40 CFR Parts 141 and 142. Drinking water regulations: Maximum contaminant level goals and national primary drinking water regulations for lead and copper; final rule. Federal Register, **56**: 26460–26464.

20. WHO (2004). Copper in Drinking-water. Background document for development of WHO guidelines for drinking water quality. WHO/SDE/WSH/03.04/88.

21. Pedersen JT, Østergaard J, Rozlosnik N *et al.* (2011) Cu(II) mediates kinetically distinct, non-amyloidogenic aggregation of amyloid-beta peptides. *J. Biol. Chem.* **286**: 26952–26963.

22. Huang X, Moir RD, Tanzi RE, Bush AI, Rogers JT. (2004) Redox-Active Metals, Oxidative Stress, and Alzheimer's Disease Pathology. *Ann. NY Acad. Sci.* **1012**: 153–163.

23. Squitti R, Barbati G, Rossi L *et al.* (2006) Excess of nonceruloplasmin serum copper in AD correlates with MMSE, CSF [beta]-amyloid, and h-tau. *Neurology* **67**: 76–82.

24. Squitti R, Bressi F, Pasqualetti P *et al.* (2009) Longitudinal prognostic value of serum "free" copper in patients with Alzheimer disease. *Neurology* **72**: 50–55.

25. Squitti R, Ghidoni R, Scrascia F *et al.* (2011) Free copper distinguishes mild cognitive impairment subjects from healthy elderly individuals. *J. Alzheimers. Dis.* **23**(2): 239–248.

26. Bucossi S, Ventriglia M, Panetta V *et al.* (2011) Copper in Alzheimer's Disease: A meta -analysis of serum, plasma, and cerebrospinal fluid studies. *J. Alzheimers. Dis.* **24**: 175–185.

27. Brewer GJ. (2010) Risks of copper and iron toxicity during aging in humans. *Chem. Res. Toxicol.* **23**: 319–326.

28. Gill JS, Bhagat CI. (1999) Acute copper poisoning from drinking lime cordial prepared and left overnight in an old urn. *Med. J. Aust.* **170**: 510.

29. Hoveyda N, Yates B, Bond CR *et al.* (2003) A cluster of cases of abdominal pain possibly associated with high copper levels in a private water supply. *J. Environ. Health.* **66**: 29–32.

30. Chug KS, Sharma BK, Singhal PC *et al.* (1977) Acute renal failure following copper sulfate intoxication. *Postgrad. Med. J.* **53**: 18–23.

31. Levy DA, Bens MS, Craun GF *et al.* (1998) Surveillance for waterborne-disease outbreaks — United States, 1995–1996. *MMWR CDC Surveill Summ.* **47**: 1–34.

32. Naha K, Saravu K, Shastry BA. (2012) Blue vitriol poisoning: A 10-year experience in a tertiary care hospital. *Clin. Toxicol. (Phila).* **50**: 197–201.

33. Spitalny KC, Brondum J, Vogt RL *et al.* (1984) Drinking-water-induced copper intoxication in a Vermont family. *Pediatrics.* **74**: 1103–1106.

34. Olivares. (1998) Copper in infant nutrition: Safety of World Health Organization provisional guideline value for copper content of drinking water. *J. Pediatric. Gastroent. Nutr.* **26**: 251–257.

35. Araya M, Olivares M, Pizarro M *et al.* (2004) Community-based randomized double-blind study of gastrointestinal effects and copper exposure in drinking water. *Environ. Health. Perspect.* **112**: 1068–1073.

36. Zacarías I, Yáñez CG, Araya M *et al.* (2001) Determination of the taste threshold of copper in water. *Chem. Senses.* **26**: 85–89.

37. Omur-Ozbek P, Dietrich AM. (2011) Retronasal perception and flavor thresholds of iron and copper in drinking water. *J. Water. Health.* **9**: 1–9.

38. Food and Nutrition Board, Institute of Medicine. (2001) *Dietary Reference Intakes: Vitamin A, Vitamin K, Arsenic, Boron, Chromium, Copper, Iodine, Iron, Manganese, Molybdenum, Nickel, Silicon, Vanadium, and Zinc.* National Academy Press, Washington DC.

39. Chambers A, Krewski D, Birkett N *et al.* (2010) An exposure-response curve for copper excess and deficiency. *Toxicol. Environ. Health. B. Crit. Rev.* **13**: 546–578.

40. Barnes DE, Yaffe L. (2011) The projected effect of risk factor reduction on Alzheimer's disease prevalence. *Lancet Neurol.* **10**: 819–28.

41. Schrag M, Mueller C, Oyoyo U, *et al.* (2011) Iron, zinc and copper in the Alzheimer's disease brain: a quantitative meta-analysis. Some insight on the influence of citation bias on scientific opinion. *Prog. Neurobiol.* **94**: 296–306.

42. Squitti R. (2012) Metals in Alzheimer's disease: a systemic perspective. *Front Biosci.* **17**: 451–72.

43. Environmental Health Criteria. 200. (1998) Copper. International Programme on Chemical Safety. World Health Organization, Geneva.

# CHAPTER 4

# ETHYLENE

Ethylene is one of the smallest gaseous organic compounds. It is present in our atmosphere in trace amounts with approximately three-quarters of it coming from natural sources such as emissions from vegetations and the rest from man-made activities such as automotive exhaust and industrial activities. At concentrations above 20% in air, it has narcotic and anesthetic effects. In fact, it was once used as anesthetic in the 1900s but was discontinued and replaced by other safer anesthetics because the mixture was flammable and liable to explode. Ethylene is an indispensable industrial chemical. The bulk of the ethylene supply today is derived from the processing of natural gas and petroleum. Almost half of the production worldwide is used for the manufacturing of polyethylene, from which plastic products are fabricated. Ethylene is the starting material for the manufacturing of polystyrene and polyvinyl polymers, ethoxylated detergents, and ethylene-glycol type antifreezes. Oxidation of ethylene generates ethylene oxide which is a potent gaseous sterilizer used in hospitals. Other products derived from ethylene included the fumigant, ethylene dibromide, and organic solvents such as industrial alcohols and trichloroethylene. Ethylene plays an important role in the plant kingdom. It is a natural plant hormone which regulates many aspects of the plant life cycle, including seed germination, root initiation, flower development, fruit ripening, senescence and response to stress.[6,7] It is interesting to know that ethylene is also involved in the process of leaf shedding. Controlling the level of ethylene has been employed as a mean of preventing the premature shedding of trees including Christmas trees.[8] It is known that change in the level of atmospheric ethylene especially in urban areas significantly influences the ripening speed of crops and hence the harvesting time.[9] Farmers used ethylene for the controlled ripening of some fruits and manipulate the ethylene concentration to lengthen the storage time of fruit and vegetables. The influence of ethylene on our food supply may be considered as directly influencing our well-being.[10]

The amount of ethylene normally present in the air is so low it is considered to be harmless to humans.[1] Ethylene is metabolized in the body to ethylene oxide, a carcinogen and mutagen. The metabolism of ethylene in the body is not fully understood but animal studies so far have not revealed ethylene itself to be carcinogenic. The International Agency for Research on Cancer[2] reviewed ethylene and concluded that it is not classifiable as to its carcinogenicity to humans (Group 3). However, there is concern that because ethylene is converted to ethylene oxide, it may be a human carcinogen under some conditions.[3–5]

Although ethylene is not a significant health concern for humans, it has in the past decades became the main character of a captivating story which involved the interplay of history, mythology, psychology, geology, chemistry and toxicology, and has captured the attention of the popular media. The way the ethylene story unfolds and the response of the scientific community and general public illustrates the often unexpected consequences of mixing mythology with serious scientific investigations. Historians and scientists have been intrigued by the story of the oracles of Delphi in ancient Greece. From 373 BC until 392 AD when it was shut down by emperor Theodosius, the temple of Delphi in central Greece was one of the major centers of worship. Inside the temple, in a small, secluded underground adyton, the high priestess, or Phythia, would sit on top of a bronze tripod and waited for inspiration from the god Apollo. At the same time she would inhale the vapor emanating from the warm spring beneath the temple and drank the water. In time she would enter into a state of trance and utter statements that are considered by his followers to be divinely inspired and prophetic. A belief of ancient writers was that the priestess was affected by a gas rising from cracks on the ground where she stood. The Greek historian and author Plutarch (46–120 AD), who was born near Delphi and once served as a priest in the temple, mentioned in one of his dialogues that a fragrance not unlike the sweetest and most expensive perfumes would occasionally fill the chamber where the priestess sat. The divine site was said to be discovered by a goatherd who observed that the animals trapped in the chasm were behaving abnormally. After reaching the chasm the goatherd himself was overcame by vapors and commenced to have visions. For centuries after, the mysteries of the oracles were buried with the ruins of the temple of Delphi. It was not until 1892 when a team of French archaeologists led by Theophile Homolle located and excavated the site. To their great disappointment they did not find any chasm beneath the temple, nor did they find fissures through which vapors may be emitted. It was concluded that the oracle of Delphi along with the vapor that induce divination were just a myth. Soon after Hemolle's report, Adolphe Oppè of the University of Edinburgh wrote an article warning against the unconditional acceptance of traditional account of the Oracle of Delphi. He pointed out the discrepancies in the various accounts of origin and performance of the ritual in the ancient texts and also mentioned the negative scientific evidence obtained by Homolle. Oppé was certain that there was no chasm and the mantic vapor coning out of the earth was a myth.[11–13] This was accepted by scholars for the next 70 years when the story took another turn.

In the 1980s, a geologist named Jelle Zeilinga de Boer was working for the Greek government studying active faults in relationship to the selection of sites for nuclear reactors. He noticed that in Delphi there were exposed fault lines to the East and West of the temple. The finding probably constituted valuable data to his geological project at the time and nothing else. But as was described by de Boer and colleague in the July 2003 issue of *Scientific American* and also vividly recounted by William Broad in his book *The Oracle: The Lost Secrets and Hidden Messages of Ancient Delphi*,[13] the finding took on a new meaning more than a decade later. In the early 1990s, de Boer shared the story with John Hale who as an archaeologist was intrigued because contemporary scholars of Greek classics all agreed that the chasm under the temple and the mantic gas was just a myth. The geologist–archaeologist team decided to visit the site and indeed they found two faults, one running

North–South and one running East–West that intersected directly under the sanctuary. The two faults intersected at the exact location where the temple was situated. They also found signs of a drain, indicating the presence of a spring. Further research revealed that the faults pass through deposits of bituminous limestone with petrochemical content as high as 20%. Suddenly, a theory took shape: The fault created a fissure in the rock under the temple. Gases and water impregnated with gases from deep under the ground rose to the surface through the fissures. Some of the gases may have intoxicating or hallucinating effects on humans. The next logical thing to do was to test samples of water from nearby spring for traces of petrochemicals. The chemist Chanton was recruited and he found traces of methane, ethane in the limestone deposits, and ethylene in samples of spring water in the vicinity of the temple site. While all three chemicals have in common in having high vapor pressure and therefore existed mostly in the gaseous state, ethylene stood out because it has a sweet odor. Henry Spiller, a toxicologist, now joined this multidisciplinary team. Through review of toxicological and medical literatures they confirmed that ethylene has been used in the past as anesthetic and at high enough concentration is psychoactive. In the words of Spiller[13] "*In the first stages, it produces disembodied euphoria, an altered mental status and a pleasant sensation. It is what street people would call getting high. The greater the dose, the deeper you go*". The hypothesis now seemed to have been proven: Contrary to accepted assumption, the temple at Delphi did once sit on an active fault with a spring running right under the sunken chamber. Stress in the fault created heat which vaporized petrochemicals in the soil. The priestess inhaled the hydrocarbon vapors (a mixture of methane, ethane and ethylene) became euphoric and then entered a trance state conducive to prophesying. Ethylene, with its known property as an anesthetic, was tentatively identified as the main culprit. It appeared multidisciplinary scientific research had been successfully applied to solve an ancient riddle.[14-16] The findings received immediate attention from the popular and scientific press. The team explained their finding in an article in *Scientific American*.[15] The title of an article on 14 August 2001 in *National Geographic* declared "*Delphic Oracle's Lips May Have Been Loosened by Gas Vapors*".

On October 2nd 2006, an issue of *Science on-line* magazine mused about "The Prophet of Gases." "*Delphic oracle was ancient glue-sniffer*" proclaimed the newspaper *Observer* on 3 August 2003. It is a wonder to behold that ethylene, one of the smallest gas molecule bubbling up from deep within the faults of Delphi, could have so profoundly influenced the life of ancient Greek in war and peace.

It did not take long for what appeared to be a perfect piece of scientific sleuthing to be challenged. Etiope and his team[18] was the first to challenge the ethylene theory. They agreed with Boer's theory that hydrocarbon vapor fumes released in the Delphi temple of Apollo were mantic and capable of inducing the priestesses to prophesize. However, they suggested that high concentrations of ethylene was thermodynamically impossible and that the intoxication vapors could be methane, carbon dioxide ($CO_2$) or benzene. High concentration of methane and $CO_2$ invading the chamber, which was probably poorly ventilated could reduce the amount of oxygen causing hypoxic reactions. Benzene was also a prime candidate because of its sweet smell. Up to 500 $\mu g/L$ of naturally occurring benzene has been found to release from bituminous rocks into ground water. Inhalation of ppm levels of benzene leads to shortness of breath, diminished mental alertness, impair vision flawed judgment, emotional instability, and unconsciousness.

Yet another theory was proposed by Piccardi *et al.*[19] Based on geochemical data they proposed that there was a deep pocket of gas fed by a hydrothermal system lying relatively deep under the Delphi temple. Episodic seismic activity may have disturbed the gas pocket which then proceeded to release vapor into the temple chamber. Based on geochemical data they suggested the vapor contains $CO_2$ and hydrogen sulfide ($H_2S$); $CO_2$ at high levels affects the brain causing dizziness, confusion and hearing and visual dysfunctions (see Table 1, Chapter 8). To accommodate the fact that $H_2S$ has a rotten-egg smell rather than a sweet smell as described by Plutarch, Piccardi *et al.* invoked the Homeric Hymn to Apollo in which Apollo was said to have slain a dragon to acquire the oracle. The rotting of the serpent in the chasm of Delphi may then be the source of the site's original name — in Greek, python is a dragon or serpent, while pythein is a verb meaning "to rot". The smell of $H_2S$ may be interpreted by the ancient Greek as the mantic vapor emitting from the rotten corpse of the serpent.

While Etiope and Piccardi differed with de Boer only in the constituents of the mantic vapor, Foster and Lehoux[20,21] criticized de Boer's work on several fronts. They argued that the concentration of ethylene identified by de Boer would have been insufficient to cause a trance-like state. The reported concentrations of ethylene, which was in the low nM level, is not unlike concentrations present in fumes emitted from alkylated-fuel lawnmowers and in urban traffic and there have never been reports of operators and pedestrians spontaneously entering a trance, not to mention prophesying. Foster and Lehoux were concerned about the unquestioned acceptance of the de Boer theory by the public and opined that "...*this tenuous argument was widely propagated because it appealed to essentially positive inclinations and sentiments in the science-reading public. Its conclusions appealed to these inclinations so effectively that readers did not notice the weak evidence and the circular arguments used to support the conclusion*". Foster and Lehoux's criticism was immediately countered by de Boer's team.[22,23] They argued that their theory (of ethylene or hydrocarbon vapor) was tested through the standard test of hypothesis procedure. To affirm their claim they pointed to the report of a second oracular shrine at a Temple of Apollo in Hieropolis, Turkey, which was also constructed over an active geological fault with gaseous vault for a cavern. Writing as an archaeologist, Hale[24] was more circumspective in his thoughts. In answering to critics who questioned why everyone who smelled the gas did not go into a trance, he quoted Plutarch as saying that the key to the Pythia's trance was her training and mental state. The gas may simply act as a trigger for those who were ready to be induced.

Amidst all the scientific arguments no one seems to have asked the question "how accurate were the prophecies?" A revealing story was recounted by William Broad in his book.[13] Croesus, who ascended to the throne of the Kingdom of Lydia (560 BC), was eager to attack Persia. He enquired of the prophetess of Delphi and the answer was that if he marched against Persia a mighty empire will be destroyed. He attacked and was soundly defeated by Cyrus the Great, the King of Persia. Incensed, the Lydian sent an envoy to ask why. The answer was that the prophecy had indeed been fulfilled; it was Croesus who got the interpretation wrong. A great empire had been destroyed — his own. It appeared that the faith of the Greek was not shaken because the alliance consulted the oracle again 80 years later. The oracle alluded to the use of "a wall of wood" which they interpreted to mean ships. Indeed, the Persian fleet of Xerxes was destroyed at the Battle of Salamis.[25] We should not be surprised by the equivocation of the oracles; if ethylene deep inside the adyton was indeed responsible for inducing the priestess into euphoria followed by trance-like state, then anything she uttered in such a mental condition may not have true meaning. Furthermore, what she had uttered may not have been intelligible. There was ample evidence that another priest or priestess was nearby whose duty was to interpret the utterances and then put meaning to them — an opportunity for intentional or unintentional human interpretations. Human folly was to blame for the equivocal prophecy. Ethylene, if present, was just an innocent bystander.

Do all these scholarly reports and debates play any role in advancing our understanding of the health effects of ethylene present in nature? The answer is probably very little. The very low concentrations of ethylene found in the atmosphere and in water are likely without effect on our health or our mental state unless, of course, if you are standing very close to an active seismic fault that is spilling out fumes rich in ethylene, hydrocarbons and other toxic gases in which case a wide varities of acute health effects would have occured. Nevertheless, the debate does remind us about some caveats of scientific reports and news report about scientific discovery. The primary concern is that some "breakthrough" studies may not be readily reproduced or verified. This is particularly true when science attempts to explain mythology. The ethylene theory may be correct but it is impossible to go back thousands of years to measure the ethylene in the vapor coming out of the chasm at Delphi.

Scientists and engineers who believe that science can ultimately solve all problems and explain all phenomena should be wary of the danger of assuming a state of scientific triumphalism. A prime example of such a mentality was illustrated by E. Tognotti.[26] At the height of the worldwide influenza pandemic known as "Spanish flu" in 1918, Richard Pfeiffer, a German bacteriologist, identified a bacterium which he was confident was the pathogenic agent for the disease. During that period of time in history, microbiologists had success after success linking bacteria with specific diseases, and people accepted Pfeiffer's claim as one more example of the triumph of medical science. The bacterium was thus named *Haemophilus Influenzae*. In the meantime the real culprit, an influenza virus raged on killing millions.

We should learn from history and be vigilant to claims of scientific breakthrough as they surface, be it in bacteriology, environmental health, or any other field. They need to be closely scrutinized and tested before being embraced as facts. Hasty, unqualified announcements and half-truths may sound riveting and gratifying at the time but they may eventually return to haunt us.

## References

1. OECD SIDS Organization for economic cooperation and development. Screening information dataset. Ethylene. [http://www.chem.unep.ch/irptc/sids/oecdsids/74851.pdf].

2. International Agency for Research on Cancer. (1994) *IARC Monographs on the Evaluation of Carcinogenic Risks to Humans*, Vol. 60. Some Industrial Chemicals. International Agency for Research on Cancer, Lyon, France.

3. International Agency for Research on Cancer. (2008) *IARC Monographs on the Evaluation of Carcinogenic Risks to Humans*. Vol. 97. 1,3-Butadiene, Ethylene Oxide and Vinyl Halides (Vinyl Fluoride, Vinyl Chloride). International Agency for Research on Cancer,. Lyon, France.

4. Walker VE, Wu KY, Upton PB *et al.* (2000) Biomarkers of exposure and effect as indicators of potential carcinogenic risk arising from in vivo metabolism of ethylene to ethylene oxide. *Carcinogenesis* **21**: 1661–1669.

5. Li Q, Csanády GA, Kessler W *et al.* (2011) Kinetics of ethylene and ethylene oxide in subcellular fractions of lungs and livers of male B6C3F1

mice and male Fischer 344 rats and of human livers. *Toxicol. Sci.* **123**: 184–198.

6. Abeles FB, Morgan PW, Saltveit ME Jr. (1993) *Ethylene in Plant Biology*, 2nd ed. Academic Press, NY.

7. Lin Z, Zhong S, Grierson D. (2009) Recent advances in ethylene research. *J. Exp. Botany.* **60**: 3311–3316.

8. MacDonald MT, Lada RR, Martynenko AI *et al.* (2010) Ethylene triggers needle abscission in root-detached balsam fir. *Trees — Struct. Funct.* **24**: 879–886.

9. Kader AA. (1985) Ethylene-induced senescence and physiological disorders in harvested horticultural crops. *HortScience* **20**: 54–57.

10. Wills RBH, Ku VVV, Shohet D *et al.* (1999) Importance of low ethylene levels to delay senescence of non-climacteric fruit and vegetables. *Austral. J. Expt. Agr.* **39**: 221–222.

11. T. Dempsey. (1918) *The Delphic Oracle: Its Early History, Influence and Fall.* Oxford: B.H. Blackwell.

12. Spiller HA, Hale JR, De Boer JZ. (2002) The Delphic oracle: A multidisciplinary defense of the gaseous vent theory. *J. Toxicol. Clin. Toxicol.* **40**: 189–196.

13. Broad WJ. (2006) *The Oracle: The Lost Secrets and Hidden Message of Ancient Delphi.* The Penguin Press, New York.

14. De Boer JZ, Hale JR, Chanton J. (2001) New evidence for the geological origins of the ancient Delphic oracle (Greece). *Geology* **29**: 707–710.

15. Hale JR, de Boer JZ, Chanton JP, Spiller HA. (2003) Questioning the Delphic oracle. *Sci Am.* **289**: 66–73.

16. Spiller HA, Hale JR, De Boer JZ. (2002) The Delphic oracle: A multidisciplinary defense of the gaseous vent theory. *J. Toxicol. Clin. Toxicol.* **40**: 189–196.

17. Hale JR, de Boer JZ, Chanton JP *et al.* (2003) Questioning the Delphic oracle. *Sci Am.* **289**: 66–73.

18. Etiope G, Papatheodorou G, Christodoulou D *et al.* (2006) The geological links of the ancient Delphic Oracle (Greece): A reappraisal of natural gas occurrence and origin. *Geology* **34**: 821–824.

19. Piccardi L, Monti C, Vaselli O *et al.* (2008) Scent of a myth: Tectonics, geochemistry and geomythology at Delphi (Greece). *J. Geol. Soc.* **165**: 5–18.

20. Lehoux D. (2007) Drugs and the Delphic Oracle. *Classical World* **101**(1): 41–56.

21. Foster J, Lehoux D. (2007) The Delphic oracle and the ethylene-intoxication hypothesis. *J. Clin. Toxicol.* **45**: 85–89.

22. Spiller H, de Boer J, Hale JR *et al.* (2008) Gaseous emissions at the site of the Delphic Oracle: Assessing the ancient evidence. *Clin. Toxicol. (Phila).* **46**: 487–488.

23. Foster J, Lehoux D. (2008) A mighty wind. *J. Clin. Toxicol.* **46**: 1098–1099.

24. Hale J. (2006) Mystery at Delphi: what would Apollo say about this? DIG October.

25. Gugliotta G. (2002) Faults suggest a high calling for Delphi priestesses. *J. Toxicol. Clin. Toxicol.* **40**: 197–198.

26. Tognotti E. (2003) Scientific triumphalism and learning from facts: Bacteriology and the "Spanish flu" challenge of 1918. *Soc. Hist. Med.* **16**: 97–110.

# CHAPTER 5

# LEAD

## Lead in the Environment

Lead is a heavy, bluish-gray metal belonging to Group 14 (IVA) of the periodic table. It occurred naturally in the Earth's crust, usually in combination with two or more other elements to form lead compounds. The most common lead ore are galena (lead sulfide), followed by anglesite (lead sulfate) and cerussite (lead carbonate). Lead is resistant to corrosion. It has a low melting point and is easily molded and shaped and can be combined with other metals to form alloys. It is not a particularly abundant element in the earth's crust but its ore deposits are readily accessible and widely distributed throughout the world. Metallic lead exists in nature but its occurrence is rare. When exposed to air and water, films of lead sulfate, lead oxide and lead carbonate are formed, which act as protective barriers that slow or halt corrosion of the underlying metal.[1]

Our environment is contaminated with lead mainly because of human activity. The amount of lead in the environment increased during the industrial revolution, and again significantly in the 1920s with the introduction of leaded gasoline.[2] Environmental lead has increased more than 1,000-fold over the past three centuries as a result of human activity. The greatest increase occurred between the year 1950 and 2000 with increased industrial activities and widespread use of leaded gasoline.[1] In the developed countries, lead level in the environment have declined significantly in the past few decades due to pollution control measures, various remedial and clean-up efforts, and the phasing out of leaded gasoline. In developing and under-developed countries, lead pollution is still a concern due to continued industrial and mining activities, and lack of pollution control and clean-up efforts.

## Production and Use

Lead is a by-product of copper–zinc and silver mining processes, accounting for 70% of total lead production. Ores containing mainly lead account for about 20%. The refining industries involved in recovery of lead from waste materials such as lead plates, batteries, cable covering, etc., also produced considerable amount of lead.

Lead is widely used in industries because of its exceptional physical and chemical properties: High density, high stability, ease of casting, high opacity to gamma and X-ray energies, low sound conductance, low melting point, and high corrosion resistance. Lead is used in its metal form, as alloy with other metals, and as chemical compounds. More than half of the total lead produced today is used for manufacturing lead-acid batteries for cars and other industries. Other major uses include cable sheathing, ammunition, and as ingredients in pigment, glass and ceramic products. Leaded glass is a component of cathode ray tube and computer monitors. Some dyes for grey hairs contain lead acetate. One indispensible use is as lead shields around instruments containing or emitting radiations, such as X-ray machines. A lead shield is worn by workers around radiation-emitting devices. There are also a few countries that still use leaded gasoline. Lead is found in leaded paints for special applications. Use of lead for solders, sinks, pesticides and weighs have been phased out in most countries although sporadic uses in some countries cannot be excluded.

## Source of Exposure[1–3]

*Food.* Food may be contaminated through lead-containing utensils, lead-glazed ceramic foodware or lead crystal ware. Lead-soldered food cans used to contribute significantly to lead found in food. Most countries have banned the use of lead-solder in cans but there is concern that some imported canned foods may still use leads solders.

***Drinking water.*** The amount of lead in natural water and municipal water is generally very low. However, lead can enter a local water supply from old lead service connections (pipes) or lead solder in building plumbings. Older homes are more liable of having lead service connections. Lead solders may still be in use in some countries. The amount of lead leached into drinking water will increase as water sits in pipes or if the water is very soft or very acidic.

In the U.S. homes built after 1986 are less likely to have lead pipes, fixtures and solders. However new homes are still at risk because "lead-free" plumbing may contain up to 8% lead. The most common problem is with brass or chrome-plated brass faucets and fixtures which can leach significant amounts of lead into water, especially hot water.[58] The lead contamination incident that occurred in a new building at the University of North Carolina was a typical example of the "new" problem, and it took the persistent effort of a university administrator, Carolyn Elfland, to bring this to light.[59] It was found that decorative brass used as faucet fixtures contained lead, which leached into the drinking water. Later on it was discovered that some valves that had "acceptable" limit of lead in fact had as much as 18 percent lead by weight on the inner surfaces contacting the drinking water.[59,60] When a plumbing manufacturer explained that beautiful faucets are formed with leaded brass, Carolyn Elfland responded, *"You're proud of a faucet that looks like a swan but poisons kids?"*[61] The most recent U.S. legislation mandates that after January 4, 2014, all faucets purchased will contain no more than 0.25 percent lead in relation to wetted surface.[62]

***Dust and soil.*** Dust and soil can be significant sources of exposure to lead for children. Lead levels in soil tend to be higher in cities, near roadways, around industrial sources that use or emit lead, near weapon firing ranges, or next to buildings where crumbling lead paint has fallen into the soil. Sources of lead in household dust can also come from within the home, especially older homes that contain lead-based paints. Lead can also be brought home by people who work in a lead-related industries, e.g. battery reclamation, construction, mining and smelting.

*Air.* Lead is released into the air through industrial emissions, smelters and refineries. Lead is also emitted into the air by cars and trucks that use leaded gasoline. Lead concentrations in the air have declined significantly in countries that have banned leaded gasoline. As of 2000, only three countries (Afghanistan, Myanmar and North Korea) still use leaded gasoline and nine countries still used both leaded and unleaded gasoline.[3,4]

*Lead Paint.* It has been known for over a century that lead paint is hazardous to children. A 1904 public health article by J. Lockhart Gibson was entitled *A plea for painted railings and painted walls of rooms as the source of lead poisoning amongst Queensland children.*[5] Children in particular were exposed to lead-contaminated house dust which originated largely from disintegrated lead-based paints.[6] Indoor and outdoor paints made before 1950 frequently contained lead. Lead-based paint in older homes is a serious health hazard if it is chipping or flaking, or if it is within the reach of children. Most countries have now prohibited the use of lead paint for interior purposes. However, exterior paints, artist paints and some touch-up coatings could still contain high amounts of lead. Some metal bridges may still have thick layers of leaded paints.

*Cathode ray tubes.* Flat screen display terminals using liquid crystal and plasma display panels are rapidly replacing the old, leaded glass-containing cathode ray tubes in televisions and computers. Unfortunately this resulted in the generation of a massive volume of discarded old monitors and display terminals. In China, for example, an estimated 6 million old televisions are discarded each year. A U.S. survey showed that much of the household computers and televisions were disposed of in municipal landfills and waste-to-energy facilities, thus contributing to lead contamination of the environment. Some countries imported cathode ray tubes and electronic equipment for storage, disposal and recycle purposes and the work areas, if not properly managed, may be heavily contaminated with lead and other metals.[63,64]

*Traditional medicine and cosmetics.* Some herbal and traditional medicine and supplements from countries in Asia, Africa, the Middle

East and South America contain high concentrations of lead and are liable to increase blood lead levels if ingested.[7,8]

Kohl is a traditional cosmetic. It is mainly applied around the eyes. It is popular in South Asia, Middle East and part of Africa. A major ingredient of Kohl is galena or lead sulfide. While some of the kohl preparations contain no lead, others have as much as 50% lead content.[9,10]

***Jewelry and toys.*** Some jewelry, charms and toys may contain lead. Lead poisoning may occur when these types of jewelry are swallowed or in close contact with the body.[11,12] Toys with lead components or surface coatings that contain lead are banned in many countries but old toys containing lead can still be found.

***Venetian blinds.*** Some older brands of plastic venetian blinds may contain lead which is used as a stabilizer. As the blinds aged, lead-containing particles and dust are released. Infants and children may be exposed to the lead component if blinds are placed at a height that is accessible to them,[3] or if dust particles are in contact with them.

***Arts supplies.*** The proposition that artists such as Vincent van Gogh and Goya[13,14] may have suffered from lead poisoning certainly raised public awareness to the hidden hazards of pigments. Unfortunately, short of exhuming the bodies and measuring the level of lead in their bones and teeth, there is no scientific way of verifying the claim. Lead is the basic ingredient of some essential pigments, e.g. lead carbonate in white lead and lead chromium in yellow chrome. People working with art supplies such as inks, dyes, paints and pastels, wax crayons, and even children's face paints may be at risk of exposure to lead.[1,3]

***Hobbies.*** It may be a surprise to many that their hobbies may expose them to lead. Activities such as glazing for pottery and glassware, stain glass soldering, jewelry making, floral and decorative arrangements, or target shooting may put a hobbyist at risk of lead poisoning.[1,3]

## Leaded Gasoline

In 1922, Thomas Midgley and co-workers at the General Motors laboratory discovered that adding tetraethyl lead to gasoline eliminated engine "knock" and improved the performance of car engines. Thus began the controversial era of leaded gasoline.[15] From its first marketing in 1923 to its gradual phasing out in the 1980s and complete ban in 1996, leaded gasoline was the predominant motor vehicle fuel. However, problems soon emerged with its production and use. In the first few years, there were an estimated 15 deaths and more than 300 psychotic cases among workers in the plants producing tetraethyl lead. Many health experts warned the general public and authority about the potential danger. One of the first warnings was issued by Mansfield Clark as he foresaw the emission of lead from car exhaust and its dangerous built-up in heavily travelled road and tunnels. Alice Hamilton of Harvard Medical School made the following points: "...*lead is a slow and cumulative poison and it does not usually produce striking symptoms that are easily recognized ... if it is a probable danger, shall we not say that it is going to be a widespread one?*".[16] Regrettably, after a brief moratorium in production to allow for exposure studies, tetraethyl lead was allowed to be marketed; and the world lived with its consequences for the next 60 years. The greatest increase in environmental lead occurred between the year 1950 and 2000, which coincided with increased worldwide use of leaded gasoline.[1] Exposure to higher lead level in turn resulted in increased lead in the body. Thus, it was clearly demonstrated that children exposed to high level of lead in air had higher blood lead levels.[17] In contrast, from 1976 to 1991, as the phasing out of leaded gasoline begins, the mean blood lead level of the U.S. population age 1 to 74 years dropped 78% from 12.8 $\mu$g/dL to 2.8 $\mu$g/dL.[1]

## History

Man's first exploitation of lead is lost in antiquity but it is likely to be one of the first metals to have been exploited because of its low melting point — metallic lead melts at a relatively low temperature of 327°C while lead sulfide, the major lead ore, melts at 1114°C. The earliest known example of metallic lead is a metal figure recovered from the Temple of Abydus in Upper Egypt around 4000 BC.[18] Lead was an important part of everyday life of ancient Greek society; pieces of lead were found among ruins of ancient Troy (3000 BC).

Lead was actively mined in Roman times.[19] At the same time, large amount of lead was produced as residue of silver mining. Both Xenophon (434–359 BC) and Lucretius (98–55 BC) warned that the smoke of lead mines in Attica was harmful to health.[18] Lead in the form of the sweet-tasting acetate salt was added to wine for taste. Lead-lined containers were routinely used for the preparation and storage of food and beverages. Because of its ability to inhibit microbial growth and slow spoiling, lead salt was purposely added to food and beverage. Lead was used by the Romans in plumbing to replace the more expensive copper and tin. It was said that Romans used to dip their leaded comb in acid before grooming. The soluble lead compound (probably lead acetate) slowly oxidized in air and turned grey hairs black. It was theorized that the common exposure to lead in everyday life resulted in widespread chronic lead poisoning of the Romans. It was also suggested that excessive amount of lead ingested through their food and beverages led to the behavioral and neurological abnormalities of several Roman emperors between 15 AD and 225 AD[20] and hence contributed to the fall of the Roman empire.[21]

The Bible[22] probably provided the oldest continuous references to lead. The Book of Job (19:24), which is dated by most scholar to at least 1500 BC, mentioned that lead was used as a base for writing and marking. In other parts of the Old Testaments, it was mentioned that lead was a residue of smelting (Jeremiah 6:29; Ezekiel 22:18, 20), as an exchange for merchandize (Ezekiel 27:12) and as a cover and weight (Zechariah 5:7–8).

It was generally assumed that Hippocrates (*ca.* 460 BC–370 BC) was the first to recognize lead as toxic. However, H.A, Waldron[23] pointed out that the passages in Hippocrates' original text (*Epidemics*

*VI, 25*) described the condition of a miner with a variety of ailments that did not correspond to lead poisoning. The earliest account of lead poisoning has therefore been credited to Nicander, a second Century BC poet and physician.[23,24]

In the medieval age, exposure to lead was also prevalent. The ancient practice of adding lead salts to beverages continued. Lead was often used instead of tin in the manufacture of pewter. Many medications contained lead. By the late medieval period, it became apparent that lead in wine was the cause of many maladies, so authorities in France, Spain and Germany prohibited the addition of lead to wine.[25] The royalties of Europe were particularly prone to have gout attacks, hence the term "disease of kings" or "royal gout".[26] Some historians assumed that the condition was an inherited trait; others blamed the affliction on indulgence in food and drinks. The indulgence theory is creditable in light of our present knowledge. We now know that dark meat and animal organs are rich in purines, a precursor of uric acid, and alcohol is a risk factor for gout. Furthermore, as described below there may be an interesting association between lead in drinks and gout attack.

**Saturnine Gout.** Gout, as a consequence of chronic lead poisoning, was described as "saturnine gout" by Musgrave in 1703. Lead was associated with the planet Saturn in Babylonian time and then with the Roman titan Saturn,[27] hence the term saturnine. It was suggested that the high level of lead in wine was contributing to the "epidemic" of gout in 18th and 19th century England.[28] Even today, some illicit alcohols or moonshines still contain high levels of lead.[29] In 1895, A.B. Garrod observed a relationship between ingestion of fortified ports, later shown to be heavily contaminated with lead, and the development of gouty arthritis.[30,57] Mamoud Logman-Adham reviewed the effects of lead on the kidneys and concluded that *"Chronic lead poisoning, secondary to occupational exposure or consumption of lead-contaminated illicit alcohol, may result in what has been termed lead nephropathy. The latter seems to develop when blood lead levels exceed 60 μg/dL. Renal histologic changes of chronic lead nephropathy are nonspecific and include* *interstitial fibrosis, tubular atrophy, and glomerular sclerosis. They lead to progressive and irreversible renal failure often associated with hypertension, hyperuricemia, and gout".*[31]

The proposed mechanism underlying the development of gout is that lead caused hyperuricemia by reducing the ability of the kidneys to secret uric acid. The accumulation of uric acid in the blood (hyperuricemia) in turn caused crystal deposition in the joints and precipitated gout attacks. There is considerable confusion as to what exactly the term "gout" meant in the olden days. Richard Wedeen[26] clarified the confusion by pointing out that "irregular gout" of old was a general term describing conditions that affected the stomach, the nervous system and the kidneys. The cause of "irregular gout" was attributed to debauchery and intemperance. This "gout" is to be separated from the modern "arthritic gout" which is reserved for inflammation of the joints due to deposition of uric acid crystals.

Charles V, the Holy Roman Emperor who reigned from 1516–1556 was known to suffer from intense arthritic pain. Jaume Ordi and colleagues[32] examined a finger tip of the mummified remains of Charles V and observed massive gouty tophi (stones) which were shown to contain deposits consistent of uric acid crystals. This result was taken as a confirmation that Charles V suffered from "arthritic gout", i.e. severe inflammation of the joints due to uric acid crystal deposits. It was further postulated that lead poisoning, or "saturnine gout", was the cause of this "arthritic gout" condition.[33] Unfortunately, Ordi's study was flawed because there was no comparative data from control tissues, i.e. specimens from a contemporary but non-diseased mummy. Furthermore, lead analysis was not done to establish excessive exposure.

The medical record of Benjamin Franklin, a more recent historical figure, was also revisited. The fact that Franklin suffered from a painful condition of bladder stone and gout was well documented.[34] The gout was clearly described as "gouty pain down in the foot" which suggested that it was the arthritic gout as defined by modern medicine and not the "irregular gout".[26] George Corner and Willard Goodwin[34] suggested that Franklin's bladder stones were uric acid crystals and therefore shared a common metabolic origin with his arthritic gout. This hypothesis was supported by Stanley Finger and Ian Hagemann,[35] who further proposed that the chronic hyperuricemia was due in part to subtle lead poisoning from the lead fortified wines that Franklin consumed. Other contributing factors were also mentioned, such as Franklin's genetic disposition (his brother also suffered from bladder stones), excessive consumption of meat (purines in meat metabolized to uric acid), and alcohol (a known risk factor for gout). It must be said that there are two types of bladder stones, calcium and uric acid with the former being the most common one by far. One way to test this hypothesis would be to analyze the composition of Franklin's stones to see which type they belong to and at the same time analyze his hair, nail and bone specimen for lead concentration.

Despite historical and modern data associating lead poisoning with arthritic gout, it is difficult to unequivocally establish a cause-effect relationship. Several regulatory agencies[1,18,31,36] either did not mention a possible link between lead poisoning and gout or just made a passing reference. Perhaps lead poisoning is not as prevalent nowadays, and wines are no longer contaminated with lead. However the question of whether chronic exposure to low levels of lead can contribute to the development of hyperuricemia and gout remains an intriguing one. To provide an answer, more studies are needed, e.g. to elucidate the toxic mechanism of lead on the kidneys and its relationship to hyperuricemia. Large scale epidemiology study is also needed to investigate the effects of low level exposure on the kidneys and development of gout.

## Health Effects

Lead is known first and foremost as a neurotoxin. Once it enters the body, lead affects virtually every organ and tissue including the brain, kidneys, liver, lungs, muscle, heart and reproductive organs. It is eventually eliminated in the urine or deposited in the bones and teeth, where it will stay for many years.

Health effects associated with exposure to high levels of lead include vomiting, diarrhea, central nervous system damages, kidney damages, convulsions, coma, even death. Lead poisoning may cause miscarriage in pregnant women. Acute exposure to high levels of lead is rare and usually involves workers in industrial settings and individuals intentionally or accidentally ingesting lead-containing substances.

Long-term, low level exposure to lead is the more likely scenario for the general public. The adverse effects are rarely life-threatening; the symptoms are often subtle, and the disease is difficult to diagnose. The main target for lead toxicity is the nervous system and the functions it controls. Lead exposure may also cause anemia, and weakness in fingers, wrists and ankles. Lead may cause an increase in blood pressure, particularly in middle-age and older people. At higher level, lead may damage the brain and kidneys.[1,31]

Inorganic lead is classified by the International Agency for Research on Cancer as probably carcinogenic to humans (Group 2A).[36] Lead and lead compounds are classified by the U.S. National Toxicological Program as "Reasonably anticipated to be human carcinogens".[37]

Blood lead level is used as the primary biomarker to assess a person's exposure status and potential health effects. Children are a particular group of concern because they are more liable to lead exposure and also more sensitive to the adverse effects. Table 1 summarizes the health effects on children and adults at blood levels from 10 to 150 $\mu$g/dL.

***Concern for children.*** Infants and toddlers are more vulnerable to the effects of lead than adults for several reasons: (1) they tend to play in an environment rich in soil, dust, paint chips, etc. that may be contaminated with lead, (2) they absorb a higher fraction (by body weight) of ingested lead, and (3) their developing nervous system is more susceptible to the adverse effects of lead. Children who swallowed large amounts of lead develop anemia, kidney damage, colic, muscle weaknesses and brain damages. Lower amount of lead can affect a child's mental and physical growth. Exposure in the womb, in infancy or in early childhood may cause slow mental development and cause lower intelligence later in childhood.[1,3]

In the United States, a blood lead level of 10$\mu$g/dL is considered the limit. The present guideline is that if a child's blood lead level

**Table 1.** Summary of Blood Lead Levels and Adverse Effects in Adults and Children.[38]

**Blood Lead Levels Associated with Adverse Health Effects**

| Children | Lead Concentration in Blood (μg/dL) | Adults |
|---|---|---|
| | **150** | |
| Death → | | ← Encephalopathy<br>← Nephropathy |
| | **100** | |
| Encephalopathy →<br>Nephropathy →<br>Frank Anemia →<br>Colic → | | ← Frank Anemia<br><br>← Male Reproductive Effects |
| | **50** | ← ↓Hemoglobin Synthesis and<br>Female Reproductive Effects |
| ↓ Hemoglobin Synthesis → | **40** | ← ↓Nerve Conduction Velocity |
| ↓ Vitamin D Metabolism → | **30** | ← Elevated Blood Pressure |
| | **20** | ← ↑Erythrocyte Protoporphyrin (men) |
| ↓ Nerve Conduction Velocity →<br>↑Erythrocyte Protoporphyrin →<br>↓Vitamin D Metabolism(?) → | | ← ↑Erythrocyte Protoporphyrin<br>(women) |
| Developmental Toxicity →<br>↓IQ, ↓Hearing, ↓Growth | **10** | |
| Transplacental Transfer → | | |

Note: ↑= increased function and =↓ decreased function.      Source: ATSDR, 1992

is: > 10μg/dL, intervention may be considered; > 20μg/dL, environmental investigation and remediation are warranted; > 45μg/dL, medical treatment may be necessary.

Since 1991 there has been research data indicating that infants and children may be adversely affected at blood lead levels <10μg/dL. These adverse changes are subtle and affect their physical, mental and intellectual developments. It was known that the IQ of children is inversely associated with blood lead level above 10μg/dL,[39] but recent reports indicated that the same adverse effect on IQ with blood lead level as low as 2.1μg/dL.[40] There is suggestion that the limit be lowered to 2μg/dL.[41]

***High blood lead in some countries.*** As a result of pollution controls and various regulatory initiatives, blood lead levels have been decreasing significantly in industrialized countries, although the same may not be true of underdeveloped and developing countries. Studies on blood lead levels in Chinese children showed that, in 2001–2007, about 24% of the children surveyed had blood lead levels greater than 10μg/dL. Blood lead levels were found to be highest in the industrial area, followed by the suburban and the urban areas.[42,43] The mean lead level was between 5 and 8μg/dl. For comparison, the average blood lead level in five-year old children in the United States was 2.7μg/dL in 1997. Because leaded gasoline was banned in China since 2000, the persistently high blood levels in Chinese children are likely due to industrial pollution. In Bangladesh, a 2007–08 survey of children in an urban industrial area found that 99% of the subjects studied had blood lead level > 10μg/dL, with the highest level measured at 64μg/dL. Industrial emissions and use of leaded gasoline by two-stroke engine vehicles were identified as possible sources of lead in these area.[44] Studies need to be carried out to assess health effect of chronic low level exposure such as mental and neurological development.

Uruguay provides an example of how a concerted educational and remedial effort was able to produce positive results.[45,46] Lead pollution in Uruguay first received official attention during the 2001 La Teja poisoning episode. Major sources of lead contamination were identified as originating from locations such as metallurgical industries, lead-acid battery processing, lead wire and pipe factories, metal foundries, metal recyclers, leaded gasoline (before December 2003), lead water pipes in old houses, and scrap and smelter solid wastes. A multidisciplinary effort to reduce lead poisoning was instituted, which included legislative actions, educational activities, and surveillance and screening programs. Interestingly, dogs were used as sentinel animals. As a result, blood lead levels in children decreased from an average of 9.6μg/dL in 1994 to 5.7μg/dL in 2004, and the proportion of children with blood lead level >10μg/dL decreased from 36% to 6.7%.

**Leaded gasoline abuse.** Volatile substance abuse such as gasoline sniffing is a worldwide problem. The presence of lead aggravated and complicated the toxic effects of gasoline sniffing. Cases of leaded gasoline abuse had been documented Northern Canada, Southwestern United States and the Australian Outback.[47–50] Encephalopathy, altered mental status and persistent psychosis due to tetraethyl lead are well-known complications of heavy and persistent gasoline sniffing.[48,50] Even for individuals with no signs of lead encephalopathy, subtle neurological and cognitive abnormalities may still be present.[51]

***Poisoning from other sources.*** There are many documented cases of high blood lead and lead poisoning due to ingestion of traditional medicine; two such cases are described here: (1) An infant was diagnosed with lead poisoning. He was found to be taking an herbal medicine for infants containing 7.5 mg of lead per dose portion. Analysis of 10 other brands of similar herbal medicine also detected significant amount of lead[52]; (2) A 60 year old South Asian man with diabetes and kidney problem exhibited signs of lead poisoning with kidney involvement. His blood lead level was 39$\mu$g/dL. He was found to be taking an Ayurvedic herbal medicine that had 49 mg of lead per tablet.[53]

Kohl had been implicated in childhood lead poisoning.[54] However, Zafar Mahmood[55] reviewed literature on Kohl and concluded that lead poisoning is not a concern. Some traditional hair dyes were also reported to have high level of lead.

The following case report illustrates the risk of lead poisoning from art-related activities. A Belgian woman used lead to make decorative arrangement for flowers. She was admitted to a hospital with a 6-month history of anorexia, weight loss and stomach pain. She was diagnosed with lead poisoning based on high blood lead, and symptoms of bilateral wrist (or foot) drops, abdominal cramps and anemia. A 50 kg block of lead was found in her basement.[56]

## Practical Points[1,3]

- A blood test from a certified laboratory is the only way to tell if you have been exposed to excessive amount of lead. If you have high blood lead your doctor will decide on the course of treatment based on your blood levels, symptoms and exposure history. Currently the guideline level for blood lead is 10 $\mu$g/dL although there are suggestions that it should be lower.

- If you have lead poisoning and it was determined to be related to exposure from your home environment, then all members of the household should be tested for blood lead level to find out about the extent of their exposure.
- Clean your house regularly to remove dust and particles. This is especially important for surfaces that young children might frequently touch.
- If you live in an older home, the dirt and dust may contain high level of lead due to disintegrated paint chips. Infants and children are the most easily exposed groups and also the most affected by lead.
- Leaded paints in a house should be removed by qualified workers. Do not use scrappers, sanders or heat gun to remove paint. They create dust and fumes that contain lead.
- Some venetian blinds made of PVC plastic contain lead as a stabilizer. Older blinds will release lead-containing dust as they disintegrate. They should be kept out of reach of children and infants.
- Older buildings may still have plumbing that contain lead. Newer buildings may also have faucet fixtures and plumbing that contain considerable amount of lead. Run the cold water tap when the water hasn't been used for a number of hours. Use only cold tap water for drinking, cooking and making baby formula, since hot water may contain more lead.
- Drinking water in public places including schools, playgrounds, shopping centers, should be test for lead frequently.
- If you work in a smelter, refinery or any other industry where you are exposed to high levels of lead, shower and change clothing before going home. This minimize the amount of lead transferred to the house.
- Do not keep food or beverages in lead crystal containers. Pregnant women and children should avoid drinking from crystal glasses that contain lead.

- If you live in a country that still allowed the use of leaded-gasoline, avoid the air and dust near roadways because they contain a higher level of lead.
- Some traditional and herbal medicines and cosmetics may contain high levels of lead. Find out about their lead content before using them. Infants, children and pregnant women should avoid using these medicines and cosmetics if a high lead level is suspected.
- If you practice arts and crafts, be aware that some of the materials including pigments, paints, glazing, soldering materials may contain lead. Caution should be taken during handling, storage and disposal.
- Some jewelry, charms, toys and decorative pieces may contain lead. They should be avoided by people of all ages. Small items that may be swallowed by infants and children are particularly dangerous.
- You may be exposed to lead through handling lead shots and leaded fishing weighs or inhaling fumes emitted after shooting a gun. People consuming games with lead shots embedded in the body may be at risk of lead poisoning.

## References

1. ATSDR. (2007) *Toxicological Profile for Lead.* Agency for Toxic Substances and Disease Registry, U.S. Department of Health and Human Services.
2. Lead and Human Health. (2011) It's Your Health. Health Canada. [http://www.hc-sc.gc.ca/hl-vs/alt_formats/pdf/iyh-vsv/environ/lead-plomb-eng.pdf].
3. Lead in Your Home. (2004) Canadian Mortgage and Housing Corporation and Health Canada. 2004. Cat. No NH15-168/1997E.
4. UNEP/PCFV. United Nations Environmental Programme/Partnership for Clean Fuels and Vehicles. [http://www.unep.org/transport/pcfv/corecampaigns/campaigns.asp#lead].
5. Gibson JL. (2005) A plea for painted railings and painted walls of rooms as the source of lead poisoning amongst Queensland children. 1904. *Public. Health. Rep.* **120**: 301–304.
6. Lanphear BP, Roghmann KJ. (1997) Pathways of lead exposure in urban children. *Environ. Res.* **74**: 67–73.

7. Lynch E, Braithwaite R. (2005) A review of the clinical and toxicological aspects of 'traditional' (herbal) medicines adulterated with heavy metals. *Expert. Opin. Drug. Saf.* **4**: 769–778.

8. Buettner C, Mukamal KJ, Gardiner P *et al.* (2009) Herbal supplement use and blood lead levels of United States adults. *J. Gen. Intern. Med.* **24**: 1175–1182.

9. Parry C, Eaton J. (1991) Kohl: A lead-hazardous eye makeup from the Third World to the First World. *Environ. Health. Perspect.* **94**: 121–123.

10. al-Hazzaa SA, Krahn PM. (1995) Kohl: A hazardous eyeliner. *Int. Ophthalmol.* **19**(2): 83–88.

11. Death of a child after ingestion of a metallic charm. (2006) *MMWR Morb Mortal Wkly Rep.* **55** 340–341.

12. Lead poisoning of a child associated with use of a Cambodian Amulet-New York City, 2009. (2011) *MMWR Morb Mortal Wkly Rep.* **60**: 69–71.

13. Weissman E. (2008) Vincent van Gogh (1853–90): The plumbic artist. *J. Medical Biography* **16**: 109–117.

14. Montes SJ. (2006) Goya, Fortuny, Van Gogh, Portinari: Lead poisoning in painters across three centuries. [Article in Spanish] *Rev. Clin. Esp.* **206**: 30–32.

15. Rosner D, Markowitz G. (1985) A "gift of God"? : The Public health controversy over leaded gasoline during the 1920s. *Am. J. Public Health* **75**: 344–352.

16. Needleman HL. (1997) Clamped in a straitjacket: The insertion of lead into gasoline. *Environ. Res.* **74**: 95–103.

17. Brunekeff B. (1984) The relationship between air lead and blood lead in children: A critical review. *Sci. Total. Environ.* **38**: 79–123.

18. UNEP 2010. Final review of scientific information on lead. United Nations Environmental Programme. Chemical Branch, DTIE. December 2010.

19. Hernberg S. (2000) Lead poisoning in a historical perspective. *Am. J. Ind. Heath.* **38**: 244–254.

20. Nriagu JO. (1983) Saturnine gout among Roman aristocrats. *N. Engl. J. Med.* **308**: 660–663.

21. Gilfillan SC. (1963) Lead poisoning and the fall of Rome. *J. Occup. Med.* **7**: 53–60.

22. Old Testaments. The Bible (KJV).

23. Waldron HA. (1973) Hippocrates and lead. *Lancet* 626.

24. Waldron HA. (1978) Did Hippocrates describe lead poisoning. *Lancet*: 1315.

25. Green DW. (1985) The Saturnine curse: A history of lead poisoning. *Southern Med. J.* **78**: 48–51.

26. Wedeen RP. (1984) Irregular gout: Humoral fantasy or saturnine malady. *Bull. N. Y. Acad. Med.* **60**: 969–979.

27. Brewster UC, Perazella MA. (2004) A review of chronic lead intoxication: An unrecognized cause of chronic kidney disease. *Am. J. Med. Sci.* **327**: 341–344.

28. Halla JT, Ball GV. (1982) Saturnine gout: A review of 42 patients. *Seminar in Arthritis and Rheumatism* **11** 307–314.

29. Reynolds RP, Knapp MJ, Baraf HSB *et al.* (1983) Moonshine and lead. *Arthritis Rheum.* **26**: 1057–1064.

30. Loghman-Adham M. (1997) Renal effects of environmental and occupational lead exposure. *Environ. Health. Perspect.* **105**: 928–939.

31. WHO. (1995) *Environmental Health Criteria 165. Inorganic lead. International Programme on Chemical Safety.* World Health Organisation. Geneva, Switzerland.

32. Ordi J, Alonso PL, de Zulueta J *et al.* (2006) The severe gout of Holy Roman Emperor Charles V. *N. Engl. J. Med.* **355**: 516–520.

33. Couper RTL. (2006) The Severe Gout of Emperor Charles V. *N. Engl. J. Med.* **355**: 1935–1936.

34. Corner GW, Goodwin WE. (1953) Benjamin Franklin's bladder stone. *J. Hist. Med.* 359–377.

35. Finger S, Hagemann IS. (2008) Benjamin Franklin's risk factors for gout and stones: From genes and diet to possible lead poisoning. *Proc. Am. Philosophical. Soc.* **152**: 189–206.

36. Lead and lead compounds. Organolead compounds. Supplement 7. (1987) International Agency for Research on Cancer (IARC) — Summaries and Evaluations. Lyon, France.

37. Report on Carcinogens. Twelfth Edition. (2011) U.S. Department of Health and Human Services. Public Health Service. National Toxicology Program. [http://ntp.niehs.nih.gov/ntp/roc/twelfth/roc12.pdf].

38. Meyer PA, Brown MJ, Falk H. (2008) Global approach to reducing lead exposure and poisoning. *Mutat. Res.* **659**: 166–175.

39. Needleman HL, Gatsonis CA. (1990) Low-level lead exposure and the IQ of children. A meta-analysis of modern studies. *J. Am. Med. Assoc.* **263**: 673–678.

40. Jusko TA, Henderson CR, Lanphear BP *et al.* (2008) Blood lead concentrations < 10µg/dL and child intelligence at 6 years of age. *Environ. Health. Perspect.* **16**: 243–248.

41. Binns HJ, Campbella C, Brownn MJ. (2007) Interpreting and managing blood lead levels of less than 10µg/dL in children and reducing childhood exposure to lead: Recommendations of the Centers for Disease Control and Prevention Advisory Committee on Childhood Lead Poisoning Prevention. *Pediatrics* **120**: e1285–e1298.

42. He K, Wang S, Zhang J. (2009) Blood lead level in children and its trend in China. *Sci. Total. Environ.* **407**: 3986–3993.

43. Zhang SM, Dai YH, Xie *et al.* (2009) Surveillance of childhood blood lead levels in 14 cities of China in 2004–2006. *Biomed. Environ. Sci.* **22**: 288–296.

44. Mitra AK, Haque A, Islam M *et al.* (2009) Lead Poisoning: An Alarming Public Health Problem in Bangladesh. *Int. J. Environ. Res. Public. Health.* 84–95.

45. Cousillas A, Pereira L, Alvarez C *et al.* (2008) Comparative study of blood lead levels in Uruguayan children (1994–2004). *Biol. Trace. Elem. Res.* **122**: 19–25.

46. Mañay NN, Cousillas A, Alvarez *et al.* (2008) Lead contamination in Uruguay: The "La Teja" neighbourhood case. *Rev. Environ. Contam. T.* **195**: 93–115.

47. Ross CA. (1982) Gasoline sniffing and lead encephalopathy. *Can. Med. Assoc. J.* **127**: 1195–1197.

48. Fortenberry JD. (1985) Gasoline sniffing. *Am. J. Med.* **79**: 740–744.

49. Cairney S, Maruff P, Burns C *et al.* (2002) The neurobehavioural consequences of petrol (gasoline) sniffing. *Neurosci. Biobehav. Rev.* **26**: 81–89.

50. Tenenbein M. (1997) Leaded gasoline abuse: The role of tetraethyl lead. *Hum. Exp. Toxicol.* **16**: 217–222.

51. Maruff P, Burns CB, Tyler P *et al.* (1998) Neurological and cognitive abnormalities associated with chronic petrol sniffing. *Brain* **121**: 1903–1917.

52. Chan H, Billmeier GJ Jr, Evans WE *et al.* (1977) Lead poisoning from ingestion of Chinese herbal medicine. *Clin. Toxicol.* **10**: 273–281.

53. Prakash S, Hernandez GT, Dujaili I *et al.* (2009) Lead poisoning from Ayurvedic herbal medicine in a patient with chronic kidney disease. *Nat. Rev. Nephrol.* **5**: 297–300.

54. Mojdehi GM, Gurtner J. (1996) Childhood lead poisoning through kohl. *Am. J. Public. Health.* **86**: 587–588.

55. Mahmood ZA, Zoha SM, Usmanghani K *et al.* (2009) Kohl (surma): Retrospect and prospect. *Pak. J. Pharm. Sci.* **22**(1): 107–122.

56. Dedeken P, Louw V, Vandooren AK *et al.* (2006) Plumbism or lead intoxication mimicking an abdominal tumor. *J. Gen. Intern. Med.* **21**(6): C1–C3.

57. Reynolds PP, Knapp MJ, Baraf HS *et al.* (1983) Moonshine and lead. Relationship to the pathogenesis of hyperuricemia in gout. *Arthritis. Rheum.* **26**: 1057–1064.

58. Lead in Drinking Water. US EPA. http://water.epa.gov/drink/info/lead/index.cfm. Last updated on Tuesday, March 06, 2012.

59. Elfland C, Scardina P, Edwards M. (2010) Lead contaminated potable water from brass devices in new buildings. *Journal of the American Water Works Association* **102**: 66–76.

60. Triantafyllidou S, Raetz M, Parks J, Edwards M. (2012) Understanding how brass ball valves passing certification testing can cause elevated lead in water when installed. *Water Research* **46**: 3240-3250.

61. Elfland takes on lead levels in plumbing as public health issue. University Gazette. The University of North Carolina at Chapel Hill. Vol 15, No. 19. P 10.
http://gazette.unc.edu/archives2/10nov 17/11-17-webpix/l 1-17-gaz.pdf.

62. Lead and Faucets — Questions and Answers. Massachusetts Water Resources Authority. MMWR Online. E:\Don Mak Book\Lead\ MWRA — About Lead and Faucets.mht. Updated July 12, 2012.

63. Nnoroma IC, Osibanjob O, Ogwuegbua MOC. (2011). Global disposal strategies for waste cathode ray tubes. *Resources, Conservation and Recycling* **55**: 275–290.

64. Xu Q, Li G, He W. (2012) Cathode ray tube (CRT) recycling: Current capabilities in China and research progress. *Waste Management* **32**: 1566–1574.

# CHAPTER 6

# MERCURY

## Chemistry and Use

Mercury as a well-known environmental contaminant has been thoroughly reviewed by several agencies.[1-3] Mercury has an atomic number of 80, an atomic weight of 200.6, and belongs to the Group 12 elements in the Periodic Table. It occurs naturally in the environment in three forms: (1) The elemental or metallic form is a silvery liquid, the only metal to be in such a state at room temperature and hence the name quicksilver. It is of toxicological significance that metallic mercury is volatile and releases mercury vapor which is colorless and odorless. The higher the temperature the more vapor is emitted; (2) Inorganic mercury compounds or salts are products of interaction of mercury with other elements such as chlorine, sulfur and oxygen. Mercuric chloride (calomel), for example, is a white powder with wide industrial use. Mercuric sulfide is another common compound. It is found as the red colored cinnabar in the ground and in ore deposits; and (3) Organic mercury compounds (organomercurials) which are formed from covalent bonding of metallic mercury with carbon.

Mercury is used in the electrolysis of sodium chloride to obtain chlorine and sodium hydroxide, and as a catalyst for the production of polymers. It is used in wiring and control devices, thermostats and cathode tubes. It is present in batteries and fluorescent lamps. Large amount of mercury is used by mining industry in the extraction of gold. Mercury-containing thermometers, barometers, manometers, and sphygmomanometers are largely being replaced by alcohol-containing devices or digital and thermistor based instruments.

Mercury is present in some medical and laboratory instruments, for example as switching devices, calibration standards and wavelength sources. Because of its strength and durability, mercury amalgam is used in restorative dental works.

Due to its toxicity, the use of mercury in consumer items such as skin ointment and cosmetics is prohibited in many countries. Mercury-containing paints, preservatives, bactericides, fungicides and disinfectants are also regulated or banned. Medical use of mercury compounds as diuretic, contraceptive, laxative, and antiseptic has largely been discontinued.

## Global Distribution and Cycling of Mercury

Mercury is a naturally occurring substance widely present in the earth's crust, waters and atmosphere. Because of its ability to exist in a vapor form, the way mercury cycles and distributes around the globe is very different from other metals. The global recycling of mercury involves chemical and biological reactions taking place in the atmosphere, the ocean, the fresh water system, the terrestrial system and the biota. The term biogeochemical cycling[4] was coined to represent this complicated process. Mercury present in the atmosphere, ocean and streams, and mercury mobilized from the earth's crust through erosion, geothermal activities and volcanic activities are the natural sources available for global recycling. However, with increased industrial activities in the past 200 years, anthropogenic sources such as mining, industrial activities and burning of coal, have drastically increased the total amount of mercury available for global recycling.

Global mercury recycling begins with mercury present on land and water and mercury generated from man-made sources. Some of the solid forms of mercury become a metallic vapor, which is subjected to atmospheric mixing and transportation. Atmospheric mercury is then returned to the ocean and terrestrial system in the form of wet deposits (rain and snow) and dry deposits (settling of particles and dust). Some of the deposited mercury is converted into insoluble mercury salts, while others are taken up by animals, vegetations and micro-organisms. In the fresh water and ocean, mercury taken up by the lower life forms is converted to methylmercury through a biological process (see *Methylmercury* below). Mercury that stays in the soil may enter a deeper stratum and become less available for recycling. Some of the mercury re-enter the atmosphere to complete the cycle. The ultimate sink for mercury is burial to the deep-ocean sediments, which occurs very slowly.

We may picture the Arctic as a pristine and unpolluted region but in reality the region is contaminated with pollutants such as organochlorines and mercury. The wildlife of the region including predatory birds, fish and marine mammals contain high level of mercury in their bodies. Residents of the Arctic tundra whose diet consisted of high percentage of wildlife also have higher than normal levels of mercury in their blood. Current studies[5-7] indicate that the Arctic region acts as a sink for mercury. Exactly how mercury was deposited in the region is still under intense investigation but one of the phenomena, the Arctic Mercury Depletion event (AMDE), first described by William Schroeder and colleagues in 1998,[5] may play an important role. Briefly, during the sunrise period, atmospheric mercury over the Arctic is oxidized to reactive mercury species through photocatalytic reactions with bromine and chlorine radicals originated from the sea. The mechanisms and dynamic of the deposition of mercury onto snow surfaces and subsequent enrichment of the biota are still being studied, but it is generally assumed that much of the mercury from polluted regions of the globe deposits at the Arctic in such a manner.

## Methylmercury

By far, the most common form of organic mercury present in the environment is methylmercury. It is lipid soluble and readily enters the nervous system including the brain. It is a neurotoxin and the most toxic of all the mercury compounds.[8] Mercury is naturally present in rivers, lakes and oceans, and mercury deposited from the atmosphere is taken up by lower life forms. Fish and shellfish consumed these lower life forms and are in turn consumed by other predatory fish. Marine animals higher up in the food chain can thus accumulate very high concentration of methylmercury through the processes of bioaccumulation and biomagnifications.[a] Methylmercury concentration in predatory fish can be elevated relative to water by a factor of a million fold. It will be impossible to include all the fish that are high in methylmercury but the list of fish in Canada and the U.S. that are high or very high in methylmercury may serve as a general guide. The list includes shark, swordfish, tuna, marlin, orange roughy, Gulf of Mexico tilefish, escolar, king mackerel, grouper, fresh water pike, fresh water bass and fresh water walleye.[9,10]

## Ancient History

Mercury sulfide or cinnabar is a mineral present in earth's crust. Because of its red coloration it was used as a red pigment since pre-historic time. There is archaeological evidence that cinnabar was used as the red pigment in paints and oracle bones in Shang Dynasty, China (1500 BC).[11] Study using modern analytical methods confirmed that cinnabar was present in colored lacquer excavated from tombs of the Warring State Period (481–221 BC) of China,[12] and in the spinning tools belonging to a young Roman woman (~100 AD).[13] The well-known Alamadèn mine in Spain was probably one of the longest running cinnabar mine in European history. It was exploited by the Carthaginians before it was taken over by the Roman following the second Punic War (218–201 BC). In the mid-sixteenth century, tons of mercury was shipped from Spain to the New World for the extraction of gold — an example of

---

[a]Bioaccumulation — progressive increase in the amount of a substance in an organism or part of an organism which occurs because the rate of intake exceeds the organism's ability to remove the substance from the body; Biomagnification — sequence of processes in an ecosystem by which higher concentrations are attained in organisms at higher trophic levels (at higher levels in the food web); at its simplest, a process leading to a higher concentration of a substance in an organism than in its food. [IUPAC glossary of terms used in toxicology, 2nd edition. *Pure Appl. Chem.* **79** (2007) 1153–1344.]

cross-continent mercury pollution initiated by man. Mercury was inseparable from alchemy,[14] a process relying heavily on the interaction of mercury with sulfur and other mineral salts. Alchemy probably begins in China and India around second and third century AD. In Indian Ayurvedic medicine, the word pertaining to alchemy is Rasa siddhi, which means processing and use of mercury and other metals for healing purpose. Paracelsus (1493–1541) the great physician and alchemist of the Renaissance period advanced the idea that the universe is constituted from three spiritual substances: the *tria prima* of mercury, sulfur, and salt. He was an ardent advocate of using mercury in medicine.[14]

Mercury was found in pigments recovered from a tomb in the Greek island of Syra dated to approximately 2000 BC. Cinnabar was one of the red pigments used in Greece during the fifth century BC.[14] Cinnabar was also confirmed by sophisticated spectrometry to be present in paint and varnishes found in a 13th century Coptic-Byzantine icon.[15] The pigment vermilion whose main ingredient is refined cinnabar was a main ingredient on the palette of European painters at least since AD 800.[16]

## Sources of Exposure — Past and Present

*Mercury in cosmetics.* The habit of applying mercury-containing preparations on the skin for bleaching purposes has been practice for centuries. Mercury in the cream seems to inhibit pigment formation and hence lightening the skin. The creams have been recognized as a cause of chronic mercury poisoning and kidney diseases.[17] Despite the well-known hazard, mercury-containing creams are still in use all over the world.[18–20] The European Union has banned the use of mercury in cosmetics and yet illegally imported mercury-containing creams are still available. In the United Kingdom, cases of chronic mercury poisoning from the use of illegal skin lightening creams are still being reported.[21] Kidney problem remains the most frequently reported hazard.[21–23] A recent case involved a 4-year-old who was mistakenly exposed to a skin-whitening cream. She was diagnosed with a brain lesion which resolved after four months.[24]

***Mercury in medications.*** Mercury is commonly found as an active ingredient in ancient medications such as Chinese herbal medicine[25] and Indian Ayurvedic medicine.[26] These types of medicinal products are not only available over the counter but now may be purchased through the internet.[27] Although many countries warned about the hazard, reports of mercury poisoning following self-medication with traditional or herbal medicines persisted.[28,29] Of grave concern is that infants and children are often the innocent victims.[30]

Mercury-containing medication was a treatment of choice for syphilis from medieval time up to middle of the 19th century. During the period of the Crusades (1096–1270 AD) many of the crusaders suffered from what was called leprosy and they were reported to be cured by a mercury-based medicine.[14] Until the early 20th century, calomel (mercurous chloride) was still used in some countries as disinfectant, as treatment for syphilis, and as a laxative.

During the American Civil War, surgeons relied on strong mercury purgatives (laxatives), such as calomel or "blue pill" (mercury and chalk) to treat diseased soldiers. Such uses often caused more problems than the diseases, with large doses causing intestinal problems, renal failure, dementia and personality changes. Soldiers were said to fear not only their diseases but also the surgeon's medications. Calomel became a remedy for any type of disease, and its overuse led the Medical Department to remove it from the pharmacopoeia in May 1863.[31] According to Norbert Hirschhorn and colleagues,[32] Abraham Lincoln's not infrequent display of anger and mood change may be attributed in part to his fondness for the "blue pill". The pill was prescribed, ironically, for melancholy or "hypochondriasis". It was said that he was astute enough to recognize the effects of the pills and stop taking it soon after his inauguration.

In the early 19th century Europe, mercury-containing candy, Wurmschokolade, was often given to children as a cure for parasitic infections and many of them came down with mercury poisoning. In the U.S., the use of calomel in medicine was banned by the Food and Drug Administration in 1960 and the frequency of mercury poisoning dropped drastically.[33]

***Teething powder and Pink disease.*** In the latter part of the 19th century doctors were baffled by a disease afflicting infants and young children that was characterized by burning palms and sole, peeling skin and anorexia. The mortality rate at the time was estimated to be approximately 7%. According to Ann Dally[34] the disease was coined the term "pink disease" by Charles Clubbe of Melbourne, Australia, because of the color of the rash. Soon doctors in Europe and North America begin to document a similar disease. For the next 70 years there were a variety of theories about the cause of pink disease. An early theory was that the disease was "either a passing bacteria infection... or else a prolonged mild infection". With the discovery of the virus as a pathogen in the 1880s, a viral origin of the disease was advanced by some clinicians. With increasing knowledge about the role of nutrition in diseases in the 1920s, nutritional deficiency was proposed as the cause of pink disease because it shared some skin symptoms with pellagra, a disease of deficiency in niacin (vitamin $B_3$). Surprisingly, no one related the disease to teething powder which was used to treat infants with teething problems at the time. Nearly half a century lapsed before a breakthrough came. In 1930, Arau suggested mercury hypersensitivity as the cause of children who developed severe skin problem following consumption of calomel medications. Arau's observation was collaborated by Fanconi in 1947. Confirmation came when Josef Warkany and Donald Hubbard[35] showed that children with acrodynia (North American term for pink disease) had high concentration of mercury in urine. The causative role of mercury was accepted, albeit slowly, by the medical community and the use of mercury in medication was discouraged or banned starting in the 1950s. What is the impact of the ban? In Britain, there were 174 death in the four years prior to 1954 when teething powders containing mercury were withdrawn from the market; in the four years following 1954 there were only 15.[36] Hindsight is always 20/20, but it is heart-wrenching to think of the children suffering from the disease while experts were grappling fruitlessly about the cause.

***Mercury in industry.*** The technique of using mercury nitrate to process fur into felt for hat making was well-guarded by the French in the 17th century. The secret became known in England when the Huguenots flee across the Channel as a result of the Edict of Nantes. What also crossed the channel along with the secret was the Hatter' Disease. As early as 1805, John Pearson described the principle psychotic feature of the disease that included excessive timidity, diffidence, shyness, and explosive loss of temper when criticized.[37] It was said that the description of the "Mad Hatter" in Alice's *Adventures in Wonderland* was based on Lewis Carroll's impression of the aberrant behavior of hat makers of that period. According to Wedeen,[38] the disease "Hatters' Shake", was first described in 1860 by J. Addison Freeman in New Jersey where a booming felt hat industry was located. He described the symptoms as tremors of the upper extremities or a shaking palsy, ulceration of the gums, loosening of the teeth, abnormal saliva, and bad breath, and ascribed the cause to the inhalation of air impregnated with mercury vapor. Unfortunately, inactions from health professionals, government, industry and the workers guild meant very little remedial actions were taken. The Second World War resolved the problem when large amount of mercury was used for detonator and the supply became scarce.[14] Change in fashion may also have accelerated the demise of the felt hat industry.

With the rising price of gold, more and more small-scale and artisanal gold mines are in operation world-wide. The estimated 15 million artisanal miners used mercury to extract gold.[39] There is concern that mercury contamination may affect the health of the miners and those living in the vicinity of the gold mines. A study showed that small-scale gold miners in Indonesia have high mercury levels in the urine and exhibited symptoms of mercury intoxication that included ataxia, tremor and movement disorders.[40] In Mongolia, female artisanal gold miners were found to have very high mercury in the urine and the inhabitant of the mining areas also had elevated mercury.[41] In Tanzania, the amalgam-burners among the artisanal gold miners were found to have mercury levels above the toxicological threshold and many of them showed signs of mercury intoxication.[42] These studies left no doubt that artisan gold miners are at high risk of mercury poisoning.

***Non-Medical Exposures.*** Elemental mercury is used in some cultural groups for magico-religious purpose.[43-45] Mercury is added to candles, incense and bath water, sprinkled or mopped on the floor of homes. The purpose is to rid the house of evil influences or to commit a ritualistic act. A study of young children living in Chicago communities with history of ritualistic use of mercury did not show elevated mercury in urine, indicating that high mercury exposure may not be a problem in this study population.[46] On the other hand, Gary Garetano and colleagues[47] measured indoor mercury vapor in a community where mercury is used for cultural purposes and identified some point sources with very high levels.

Poisoning by mercury vapor at home is a common and worldwide problem to the present day. Mercury spilled on the floor and rugs can produce enough mercury vapor to poison the occupants.[48,49] Worst still, attempts to remove the mercury by vacuuming in fact creates more aerosol and exposes occupants to higher level of mercury vapor.[50,51] Children were at higher risk of exposure because they spent more time on the floor and also breathed in more ground level aerosol and contaminants. Children are also more vulnerable to the effects of mercury because of their rapid growth and development.[52] A communicate from the U.S. Center for Diseases Control and Prevention[53] alerted the public to the fact that mercury poisoning can occur as a result of moving to an already contaminated house.

***Compact fluorescent lamps.*** Because of its energy-saving attribute and long lifespan, compact fluorescent lamps (CFL) are widely used in business and homes. The lamps contain small but variable amounts of mercury (0.1–10 mg/lamp) as an essential ingredient in the light emission process. Although the mercury is located in a sealed compartment, it nevertheless constitutes a potential source of human exposure and, in the long run, contributes to the contamination of the environment. On the other hand, argument has been put forward that the lower energy consumption leads to reduced amounts of coal being burned in power plants, which in turn lessen the amounts of mercury released into the environment.[76-78] Recent studies indicated that broken CFL continuously release mercury vapor for weeks and even months. If the rooms are poorly ventilated, the accumulated mercury may exceed the safe human exposure limit of 0.2 µg/m.[1] Many regulatory agencies now have detail guidance on how to avoid exposure to mercury vapor from CFL.[76,79] Another grave concern is that CFL contribute substantially to the total mercury inputs in municipal solid wastes, thus further contaminating the ground and the air. There are calls for more stringent disposal and recycling processes for CFL.[77,80]

## Mass Scale Mercury Poisoning

*Minamata Disease.* Half a century after the outbreak of Minamata disease, Masazumi Harada gave a retrospective account of the incident.[54] Early signs of problems were noted in 1953 at Minamata, a fishing village in the Southern tip of the Japanese Archipelago; fish and cats were dying, seabirds felt from the sky. By 1956 an outbreak of a disease of the central nervous system was recognized in the residents of Minamata. The disease symptoms included convulsion, paralysis of the extremities, impaired speech and eyesight, and many sufferers eventually died. Some patients showed acute symptoms, other had chronic diseases that appeared several years later. The cause of the disease was determined to be methylmercury poisoning. It was found that a chemical company had discharged 600 tons of methylmercury into the Minamata Bay. As a result the local fish and shellfish were contaminated. When people consumed the contaminated seafood, methylmercury accumulated in the body with the brain being the target organ. A few years later, it was found that children born to mothers in Minamata were also affected with neurological disorders that included cerebral palsy and mental retardation.[55] Methylmercury was found to pass through the placenta and cause damage to the fetus. A second outbreak of Minamata disease occurred in Niigata District of Honshu Island in 1965. It was then that serious efforts were made to control the sources of contamination. As of January 1993, 2255 people have been diagnosed with Minanmata disease. In addition about 10,000 persons are suspected of the disease. The reaction of the government and the industry was painfully slow, which prompted a Japanese commentator, as quoted by D'Itri,[56] to equate their behavior to the disease itself: lack of vision, loss of coordination and sensation and (when faced with the problem) spasmodic convulsion.

*Iraq methylmercury poisoning.* Methylmercury poisoning occurred in Iraq in a massive scale in the winter of 1971–1972. The cause was the consumption of methymercury-treated seed grain. Of the approximately 50,000 people exposed, 6530 were admitted to hospitals with poisoning of which 459 died.[57] As with the Minamata disease, damage to the central nervous system was the major symptom. Severe brain damage was also seen in newborn due to prenatal exposure.[58] This tragic incidence could have been prevented because mercury fungicide had already been identified as the culprit in earlier poisoning: Iraq in 1955–1956, 1959–1960, Pakistan in 1961, and Guatemala in 1965. The use of methylmercury as a fungicide is now discouraged worldwide and discontinued in some countries.[8]

## Epidemiological Studies

Following the Minamata and Iraq poisoning incidents, there was great concern about the safety of food that contained methylmercury. Several teams of scientists conducted clinical and epidemiological studies on humans consuming large quantities of seafood which is naturally rich in methylmercury. Gary Myers' team from the University of Rochester is one of the most active.[57] After participating in the investigation of the Iraq poisoning incident in the 1970s they focused their attention on adult population in Samoa whose diet consisted of large amounts of fish. They found no neurological indication of methylmercury poisoning in the population.[59] Next, two groups of adult Peruvians were studied, one from coastal region that consumed large amount of seafood and the other from inland consuming only small amount. They found no association between reported paresthesias (numbness and pin-pricking of the extremities), an early sign of neurological disorder, with methylmercury exposure.[60] From 1980s until the present times, the team studied the effect of methylmercury on child development in Seychelles, an island state in the Indian Ocean. They found very few associations between parental methylmercury exposure and the children's developmental and behavioral outcomes. Some indicators were positive while others were negative and the pattern was inconsistent.[61,81] Recently, they have observed an adverse association between a measure of scholastic achievement with postnatal methylmercury exposure, but only for males.[62] They suggested that further studies that include education factors are needed to confirm the result.

While the Seychelles study was ongoing, other children studies reported positive results. A study on 6- and 7-year-old New Zealand children reported an adverse effect of prenatal mercury exposure on scholastic and psychological test scores.[63] Studies conducted in Faroe Islands reported adverse associations between prenatal methylmercury and tests of memory, attention, language, motor function, and visual spatial perception.[64,65] Further study showed that the association of neurobehavioral deficit with prenatal exposure to methylmercury from seafood diet was much stronger than with the co-contaminant, polychlorinated biphenyls.[81] A study on the cognitive development of pre-school children of Granada also showed an adverse association with mercury exposure.[66] The brief description above does not do justice to the large amount of research work performed nor conveys the complexity involved in measuring and interpreting the subtle effects. However the conflicting results between the Seychelles studies and other studies underscore the difficulties involved in conducting human studies when exposure level is often only slightly higher than normal and changes, if present, are often subtle. Confounding factor is also a problem. For example, there is suggestion that beneficial nutrient factors such as essential fatty acids in fish may have influenced the test results.[82] For these reasons testing with animals is indispensable because it circumvents some of the problems faced by human studies and provide broader and more definitive dose-response results.

## Dental Amalgam

Safety of mercury-based dental amalgams must be one of the most hotly debated and controversial topic related to mercury. Despite the endless debates and concerns, mercury amalgam is still used in dentistry albeit with much care and control.[67] A brief history of the evolution of the issue is presented here followed by a summary of some current recommendations. Some alternate views will also be summarized to reflect the diversity of opinion.

The use of dental filling, some of which contained mercury, dated back centuries in China and Europe.[67] Modern mercury amalgam was developed in 1818 by Louis Nicolas Regnart, a French physician. His formulation, unlike older metallic fillings, did not require heating and was therefore easier to use. Today's dental amalgam is formed from approximately equal parts of liquid mercury and a metallic alloy consisting of silver, copper and tin. Controversy about dental amalgam is not new.[67,68] As soon as the amalgam was introduced to America in 1833, it was boycotted by dentists but not for medical reasons. They were simply too accustomed to the traditional restoration materials, primarily gold and tin. In the 1880s, in the midst of the debate about amalgam, E.S. Talbot produced some of the earliest scientific evidence that mercury was released by dental amalgam. He cited several cases of illness that were attributed to mercury poisoning. In the 1920s and 1930s, a German chemist Alfred Stock again raised concerns about mercury vapor in amalgam. Unfortunately, further investigations were disrupted by the advent of the Second World War. Modern concern arises in the 1970s and 80s, when a series of studies demonstrated mercury vapor was in the breath of subjects with dental amalgams.[67] Furthermore, chewing increased the amount of mercury vapor emitted. Animal studies showed that mercury from amalgam accumulated in the organs and tissues of the bearer. These findings stirred public concerns about possible health effect of dental amalgams and stimulated more animal and human studies. Data from these studies have been analyzed and reviewed continuously by professional associations, government agencies, and public interest groups. The current and predominant opinion is that, with some qualifications and restrictions, dental amalgam can be used. Table 1 summarized the recommendations/conclusions of (1) Health Canada[69] (2) the Scientific Committee on Emerging and Newly Identified Health Risk (SCENIHR)[70] in its submission to the EU commission, and (3) U.S. Food and Drug Administration.[71] These conclusions were not met with unanimous

**Table 1.** Recommendation of Health Canada, U.S. Food and Drug Administration (U.S. FDA), and Scientific Committee on Emerging and Newly Identified Health Risk (SCENIHR) on Dental Amalgam.

| Health Canada (1996)[9] | SCENIHR 2008[70] | U.S. FDA 2009[71] |
|---|---|---|
| 1. Dental amalgam contributes detectable amounts of mercury to the body, and is the largest single source of mercury exposure for average Canadians. However, this exposure is not causing illness in the general population. | ....It is generally concluded that no increased risks on adverse systemic effects exist and we do not consider that the current use of dental amalgam poses a risk of systemic disease. It is recognized that some local adverse effects are occasionally seen with dental amalgam fillings, but the incidence is low and normally readily managed. It is also recognized that there have been some report of reactions to dental amalgam, which are not supported by scientific evidence, but indicate that very occasionally an individual may have unexplained atypical physical or other reactions attributed to mercury. | 1. FDA considers dental amalgam fillings safe for adults and children ages six and above. |
| 2. Current evidence does not indicate that mercury contributes to Alzheimer's Disease, Amyotrophic Lateral Sclerosis, Multiple Sclerosis or Parkinson's Disease. | | 2. FDA concludes that the existing data support a finding that infants are not at risk for adverse health effects from the breast milk of women exposed to mercury vapor from dental amalgam. |
| 3. It is advisable to avoid procedures involving amalgam in pregnant women or individuals with kidney disease. | | 3. The estimated daily dose of mercury vapor in children under age six with dental amalgams is also expected to be at or below levels that the EPA and the Centers for Disease Control and Prevention (CDC) consider safe. |
| 4. It is prudent to reduce human exposure to mercury where safe and practical alternatives exist. | | 4. Pregnant or nursing mothers and parents with young children should talk with their dentists if they have concerns about dental amalgam. |
| 5. Evidence does not warrant the removal of existing amalgam fillings from individuals who have no indications of adverse effects. | | 5. If you are allergic to any of the metals in dental amalgam, you should not get amalgam fillings. |

acceptance. There has been, for example, a court case challenging the U.S. FDA's classification of dental amalgam.[72] The SCENIHR conclusion[70] was challenged by Joachim Mutter[73] on points related to total body burden, toxic tissue levels, correlations between mercury levels in body fluids and tissues, half-life of mercury in tissues, synergistic action of mercury with other metals and methodological issues. Active research is still going on. For example, Gene Watson *et al.*[74] recently studied 587 mother-child pairs enrolled in the Seychelles project and found no support for the hypothesis that prenatal mercury exposure arising from maternal dental amalgam restorations results in neurobehavioral consequences in the child.

In the mean time, several countries including Norway, Sweden, Germany, and Denmark have taken steps to regulate or restrict the use of dental amalgam.[82] There is at least one positive outcome from the ongoing debate, the dental professionals are unanimous about the risk of mercury vapor as an occupational hazard and have made significant progress in specifications for a safe working environment.[75]

## Practical Points

- Methylmercury is a neurotoxin. A common exposure route is through ingestion of seafood rich in methylmercury. The most vulnerable population is women who are pregnant or might become pregnant, breastfeeding mothers and young children.
- Many countries provide advisories on fish that are high in mercury and the consumption limits. Watch for local advisories on fish and seafood for levels of methylmercury. Do not abstain fish and seafood because they are an important part of a healthy balanced diet.
- Pregnant women, nursing mothers and parents with young children should consult their dentists about the most appropriate filling materials. Some countries already have advisories or regulations on dental amalgam.
- If you spill mercury, contact your local authority for advice on cleaning and disposal. Never try to clean the spilled area with vacuum cleaner; it will only raise the mercury level in the air.

- Broken fluorescent lamps should be cleaned up using stiff papers, sticky tapes and wet towels. Never use a vacuum cleaner for clean up. Place cleaned up materials in a sealed plastic bag or glass jar with a metal lid for special disposal. The room should be well ventilated during and after the clean up.[76]
- Do not let your children play with mercury or mercury containing devices such as calomel batteries, fluorescent light bulbs, and thermometers.
- Avoid herbal, traditional and patent medications that contain mercury. Some skin-lightening-cream also contains high level of mercury.
- Handle products containing calomel (mercury chloride) with caution.
- If you suspect mercury poisoning, consult a doctor. A positive urinary and/or blood test combined with symptoms and a history of mercury exposure would confirm mercury poisoning.

## References

1. Agency for Toxic Substances and Disease Registry. (1999) *Toxicological Profile for Mercury*, U.S. Department of Health and Human Services, Public Health Service, Agency for Toxic Substances and Disease Registry, Atlanta, Georgia. Report.
2. World Health Organization. (1993) Beryllium, Cadmium, Mercury, and Exposures in the Glass Manufacturing Industry. In *International Agency for Research on Cancer Monographs on the Evaluation of Carcinogenic risk to Humans*, Vol. 58. World Health Organization, Lyon, France.
3. World Health Organization. (1991) *Environmental Health Criteria 118. Inorganic mercury, International Programme on Chemical Safety*. World Health Organization, Geneva.
4. Selin NE. (2009) Global biogeochemical cycling of mercury: A review. *Annu. Rev. Environ. Resour.* **34**: 43–63.
5. Schroeder WH, Anlauf KG, Barrie LA *et al.* (1998) Arctic springtime depletion of mercury. *Nature* **394**: 331–332.
6. Lindberg S, Brook S, Lin CJ *et al.* (2002) Dynamic oxidation of gaseous mercury in the Arctic troposphere at Polar Sunrise. *Environ. Sci. Technol.* **36**: 1245–1256.

7. Poissant L, Zhang HH, Canário J, Constant P *et al.* (2008) Critical review of mercury fates and contamination in the arctic tundra ecosystem. *Sci. Total Environ.* **400**: 173–211.

8. World Health Organization. (1990) *Environmental Health Criteria 101. Methylmercury. International Programme on Chemical Safety.* World Health Organization, Geneva.

9. The Safety of Dental Amalgam, Health Canada. [http://www.hc-sc.gc.ca/dhp-mps/alt_formats/hpfb-dgpsa/pdf/md-im/dent_amalgam-eng.pdf].

10. Groth E. (2010) Ranking the contributions of commercial fish and shellfish varieties to mercury exposure in the United States: Implications for risk communication. *Environ. Res.* **110**: 226–236.

11. Britton BS. (1937) Oracle bone color pigments. *Harvard J. Asiaitc. Stud.* **11**: 1–3.

12. Wei S, Pintus V, Pitthard V *et al.* (2011) Analytical characterization of lacquer objects excavated from a Chu tomb in China. *J. Archaeol. Sci.* **38**: 2667–2674.

13. Aurisicchio C, Ferro D, Martinelli G *et al.* (2002) A study of a distaff of the second century A.D. from a necropolis of Boccone D'Aste (Roma, Italy) — tomb 75. *J. Cult. Herit.* **3**: 107–116.

14. Goldwater, Lenard J. (1972) *Mercury. A History of Quicksilver.* York Press, Baltimore, Maryland.

15. Abdel-Ghani M, Edwards HGM, Stern B, Janaway R *et al.* (2009) Characterization of paint and varnish on a medieval Coptic-Byzantine icon: Novel usage of dammar resin. *Spec. Acta A Mol. Biomol. Spec.* **73**: 566–575.

16. Gettens RJ, Feller RL, Chase WT. (1972) Vermillion and cinnabar. *Stud. Conserv.* **17**: 45–69.

17. Al-Saleh I, Del-Doush I, Shinwari N, Al-Baradei R *et al.* (2008) Does low mercury containing skin-lightening cream (Fair and Lovely) affect the kidney, liver and brain of female mice? *Cutan. Ocul. Toxicol.* **22**: 281–302.

18. Otto M, Ahlemeyer C, Tasche H, von Mühlendahl KE *et al.* (1994) Mercury exposure. *Nature* **367**: 110.

19. Sin KW, Tsang HF. (2003) Large-scale mercury exposure due to a cream cosmetic: Community-wide case series. *Hong Kong Med. J* **9**: 329–334.

20. Nakano Glenn E. (2008) Yearning for lightness: Transitional circuits in the marketing and consumption of skin lighteners. *Gender Soc.* **22**: 281–302.

21. Choudhury K, Morris J, Harrison H, O'Moore E *et al.* (2011) Dangers from mercury. *British Med. J.* **342**. In Press.

22. Oliveira DB, Foster G, Savill J *et al.* (1987) Membranous nephropathy caused by mercury-containing skin lightening cream. *Postgrad. Med. J.* **63**: 303–304.

23. Li S-J, Zhang S-H, Zheng C-X *et al.* (2010) Mercury-induced membranous nephropathy: Clinical and pathological features. *Clin. J. Am. Soc. Nephrol.* **5**: 439–444.

24. Benz MR, Lee SH, Kellner L *et al.* (2011) Hyperintense lesions in brain MRI after exposure to a mercuric chloride-containing skin whitening cream. *Eur. J. Pediatr.* **170**: 747–750.

25. Ernst E. (2002) Toxic heavy metals and undeclared drugs in Asian herbal medicine. *Trends Pharmacol. Sc.* **23**: 136–139.

26. Saper RB, Kales SN, Paquin J *et al.* (2004) Heavy metal content of Ayurveduc medicine products. *J. Am. Med. Assoc.* **292**: 2868–2873.

27. Saper, RB, Phillips RS, Sehgal A *et al.* (2008) Lead, mercury, and arsenic in US-and Indian-manufactured ayurvedic medicines sold via the internet. *J. Am. Med. Assoc.* **300**: 915–923.

28. Koh C, Kwong KL, Wong SN. (2009) Mercury poisoning: A rare but treatable case of failure to thrive and developmental regression in an infant. *Hong Kong Med. J.* **15**: 61–64.

29. Li SJ, Zhang SH, Chen HP, Zeng CH *et al.* (2010) Mercury-induced membranous nephropathy: Clinical and pathological features. *Clin. J. Am. Soc. Nephrol.* **5**: 439–444.

30. Counter SA, Buchanan LH. (2004) Mercury exposure in children: A review. *Toxicol. Appl. Pharmacol.* **198**: 209–230.

31. Kohl RM. (2004) This godforsaken town: Death and disease at Helena, Arkansas, 1862–1863. *Civil War Hist.* **50**(2): 109–144.

32. Hirschhorn N, Feldman RG, Greaves IA. (2001) Abraham Lincoln's blue pills: Did our 16th president suffer from mercury poisoning? *Perspect Biol. Med.* **44**: 315–332.

33. Black J. (1999) The puzzle of pink disease. *J. R. Soc. Med.* **92**: 478–481.

34. Dally A. (1997) The rise and fall of pink disease. *Soc. Hist. Med.* **10**: 291–304.

35. Warkany J, Hubbard DM. (1948) Mercury in the urine of children with acrodynia. *Lancet* **1**: 829.

36. Danthan JG, Harvey CC. (1965) Pink disease–ten years after (the epilogue). *Br. Med. J.* **1**: 1181–1182.

37. Waldron HA. (1983) Did the mad hatter have mercury poisoning? *Brit. Med. J.* **287**: 1961.

38. Wedeen RP. (1989) Were the hatters in New Jersey mad? *Am. J. Ind. Med.* **16**: 225–233.

39. Spiegel SJ, Yassi A, Spiegel JM, Veiga MM *et al.* (2005) Reducing mercury and responding to the global gold rush. *Lancet* **366**: 2070–2072.

40. Bose-O'Reilly S, Drasch G, Beinhoff C *et al.* (2010) Health assessment of artisanal gold miners in Indonesia. *Sci. Total Environ.* **408**: 713–725.

41. Steckling N, Bose-O'Reilly S, Gradel C *et al.* (2011) Mercury exposure in female artisanal small-scale miners (ASGM) in Mongolia: An analysis of human biomonitoring (HBM) data from 2008. *Sci. Total Environ.* **409**: 994–1000.

42. Bose-O'Reilly S, Drasch G, Beinhoff C *et al.* (2010) Health assessment of artisanal gold miners in Tanzania. *Sci. Total Environ.* **408**(4): 796–805.

43. Wendroff AP. (1990) Domestic mercury pollution. *Nature* **347**: 623.

44. Wendroff AP. Magico-religious mercury use in Caribbean and Latino communities: Pollution, persistence, and politics. *Environ Pract.* **7**: (2005) 87–96.

45. Greenberg MI. (1999) Mercury hazards widespread in magioc-religious practices in US. *Emergency Medicine News* **XXI**: 24–25.

46. Rogers HS, McCullough J, Kieszak S *et al.* (2007) Exposure assessment of young children living in Chicago communities with historic reports of ritualistic use of mercury. *Clin. Toxicol.* **45**: 240–247.

47. Garetano G, Gochfeld M, Stern AH. (2006) Comparison of indoor mercury vapor in common areas of residential buildings with outdoor levels in a community where mercury is used for cultural purpose. *Environ. Health Perspect.* **114**: 59–62.

48. Sexton DJ, Powell KE, liddle J *et al.* (1978) A common nonoccupational outbreak of inorganic mercury vapor poisoning. *Arch. Environ. Health* **33**: 186–191.

49. McNeil NI, Issler HC, Oliver RE, Wrong OM *et al.* (1984) Domestic metallic mercury poisoning. *Lancet* 269–271.

50. Bonhomme C, Gladyszaczak-Kohler, Cadou A *et al.* (1996) Mercury poisoning by vacuum-cleaner aerosol. *Lancet* **347**: 115.

51. Rennie AC, McGregor-Schuerman M, Dale IM *et al.* (1999) Mercury poisoning after spillage at home from a sphygmomanometer on loan from hospital. *Brit. Med. J.* **319**: 366–367.

52. Bose-O'Reilly S, McCarthy KM, Stekling N, Lettmeier B *et al.* Mercury exposure and children's health. *Curr. Probl. Pediatr. Adolesc. Health Care* **40**: 186–215.

53. MMWR. (1989) Elemental mercury vapor poisoning — North Carolina, U.S. Centers for Disease Control and Prevention. *Morbidity and Mortality Weekly Report* (MMWR) **38**: 707–772.

54. Harada M. (1994) Environmental contamination and human rights–case of Minamata disease. *Ind. Environ. Crisis Q* **8**: 141–154.

55. Kondo K. (2000) Congenital Minamata disease: Warning from Japan's experience. *J. Child Neurol.* **15**: 458–464.

56. D'Itri PA, D'Itri FM. (1978) Mercury contamination: A human tragedy. *Environ. Manag.* **2**: 3–16.

57. Myers GJ, Davidson PW, Cox C *et al.* (2000) Twenty-seven years studying the human neurotoxicity of methylmercury exposure. *Environ. Res. A.* **83**: 275–285.

58. Bakir F, Rustam H, Tikriti S *et al.* (1980) Clinical and epidemiological aspects of methylmercury poisoning. *Postgrad. Med. J.* **56**: 1–10.

59. Marsh DO, Turner M, Smith JC *et al.* (1974) Methyl mercury in human populations eating large quantities of marine 7sh. II. American Samoa: Cannery workers and 7shermen. *Proc. First Int. Mercury Conf.* **2**: 235–239.

60. Turner MD, Marsh DO, Smith JC *et al.* (1980) Methylmercury in populations eating large quantities of marine fish. *Arch. Environ. Health* **35**: 367–378.

61. Myers GJ, Davidson PW, Cox C *et al.* (2003) Prenatal methylmercury exposure from ocean fish consumption in the Seychelles child development study. *Lancet* **361**: 1686–1692.

62. Davidson PW, Leste A, Benstron E *et al.* (2010) Fish consumption, mercury exposure, and their associations with scholastic achievement in the Seychlles child development study. *Neurotoxicology* **31**: 439–447.

63. Crump KS, Kjellstrom T, Shipp AM *et al.* (1998) Influence of prenatal mercury exposure upon scholastic and psychological test performance: Benchmark analysis of a New Zealand cohort. *Risk Anal.* **18**: 701–713.

64. Grandjean P, Weihe P, White RF *et al.* (1997) Cognitive deficit in 7-year old children with prenatal exposure to methylmercury. *Neurotoxicol. Teratol.* **19**: 417–428.

65. Grandjean P, Weihe P, White RF *et al.* (1998) Cognitive performance of children prenatally exposed to safe levels of methylmercury. *Environ. Res.* **77**: 165–172.

66. Freire C, Ramos R, Lopez-Espinosa M-J *et al.* (2010) Hair mercury levels, fish consumption, and cognitive development in preschool children from Granada, Spain. *Environ. Res.* **110**: 96–104.

67. Cartland Jr. RF. The US Dental Amalgam Debate, 2010 Meeting of the FDA Dental Product Panel. [http://scribd.com/33766428/Amalgam-Its-History-and-Perils].

68. Hyson Jr. JM. (2006) Amalgam-its history and peril. *J. Calif. Dent. Assoc.* **34**: 215–229.

69. The Safety of Dental Amalgam, Health Canada, Ministry of Supply and Services Canada. (1996) Cat. H49–105/1996E. [http://www.hc-sc.gc.ca/dhp-mps/md-im/applic-demande/pubs/dent_amalgam-eng.php#a8] [accessed on 17th November 2006].

70. SCENIHR. (2008) Scientific Committee on Emerging and Newly Identified Health Risk. The safety of dental amalgam and alternative dental restoration materials for patients and users. May 6. [http://ec.europa.eu/health/ph risk/risk en.htm].

71. US FDA Medical Devices, Dental Amalgam, Potential Risk. [http://www.fda.gov/MedicalDevices/ProductsandMedicalProcedures/DentalProducts/DentalAmalgam/ucm171094.htm] [accessed on 21st February 2012].

72. Edlich RF, Cross CL, Wack CA *et al.* (2008) The food and drug administration agrees to classify mercury fillings. *J. Environ. Pathol. Toxicol. Oncol.* **27**: 303–305.

73. Mutter J. (2011) Is dental amalgam safe for humans? The opinion of the Scientific Committee of the European Commission. *J. Occup. Med. Toxicol.* **6**: 2–17.

74. Watson GE, Lynch M, Myers GJ *et al.* (2011) Prenatal exposure to dental amalgam: Evidence from the Seychelles child development study main cohort. *J. Am. Dent. Assoc.* **142**: 1283–1294.

75. Duncan A, O'Reilly DS, McDonald EB *et al.* (2011) Thirty-five year review of a mercury monitoring service for Scottish dental practices. (in press).

76. Neurotoxicology and Teratology 34. (2012) What are the Connections between Mercury and CFLs? http://www.epa.gov/cfl-hg.html. Last updated on Wednesday, March 07, 2012. 466–4726.

77. Hu Y, Chen H. (2012) Mercury risk from fluorescent lamps in China: Current status and future perspective. *Environmental International* **44**: 141–150.

78. Li Y, Yin L. (2012) Environmental release of mercury from broken compact fluorescent lamps. *Environ. Engineer. Sci.* **287**: 687–692.

79. Mercury in Compact Fluorescent Lamps. (2010). The Scientific Committee on Health and Environmental Risks (SCHER). http://ec.europa.eu/health/scientific_committees/opinions_layman/mercury-in-cfl/en/mercury-cfl/about-mercury-cfl.htm#24.

80. Silveria GTR, Chang SY. (2011) Fluorescent lamp recycling initiatives in the United States and a recycling proposal based on extended producer responsibility and product stewardship concepts. *Waste Manag. Res.* **29**: 656–668.

81. Grandjean P, Weihe P, Nielsen F *et al.* (2012) Neurobehavioral deficits at age 7 Years associated with prenatal exposure to toxicants from material seafood diet. *Neurotoxicology and Teratology* **34**: 466–472.

82. Burke FJT. (2004) Amalgam to tooth-coloured materials — implications for clinical practice and dental education: Governmental restrictions and amalgam-usage survey results. *J. Density* **32**: 343–350.

# CHAPTER 7

# NITRATE

Nitrate, together with ammonia and nitrogen, are essential components of one of the most important nutrient cycles found in terrestrial ecosystems, the nitrogen cycle. The cycle begins with nitrogen fixation, a process by which micro-organisms called diazotrophs incorporate atmospheric nitrogen into readily usable form of nitrogen compounds such as ammonia and nitrate. These nitrogen compounds are ultimately incorporated into biomolecules such as DNA and proteins, which are essential for life.[1] Another natural processes of nitrogen fixation is through lightning. The energy of lightning converts nitrogen in the atmosphere to nitric oxide (NO), which reaches the ground as nitrous acid and metabolizes to nitrate by micro-organisms.[2] Salts of nitrate are readily soluble in water. Indeed nitrate is the most common chemical contaminants in groundwater aquifers.[3]

With the invention of industrial processes for nitrogen fixation in the late 19th century, nitrate became the primary source of fertilizers in agriculture. Concomitantly, the pollution of nitrate in the aquatic environment and surface water has become a serious concern.[4] Nitrate is present in large quantities in fertilizers and ammunitions, and in small quantities in products such as specialty glasses for television, computer monitors and some toothpastes. Some vegetables are rich in nitrates, and many meat products are preserved and cured in nitrate and nitrite.

## Nitrate in History

Ancient civilizations in the Middle East, China and India are known to preserve meats in salts. Saltpetre (a mineral form of potassium

111

nitrate), or "niter", was mined in ancient China and India long before the Christian era. The use of salt and saltpetre in meat curing was commonplace in mediaeval times, and the effect of saltpetre on meat color was recognized.[5] According to Klaus Lauer,[6] the use of saltpetre for the coloring and preservation of food was first documented in Germany between 1600 and 1750. The underlying mechanism by which meat is preserved by nitrate was discovered in the late 19th and early 20th centuries.[5] Nitrate is converted by bacteria into nitrite, a potent antibacterial agent that inhibits the growth of harmful bacteria commonly involved in food processing such as *Clostridium botulinum*, *Salmonella* and *Esherichia coli*.

Potassium nitrate, together with charcoal powder and sulfur, were the main ingredients of gunpowder when it was first formulated by the Chinese in the seventh century. The popular invention was used to prepare fireworks and an assortment of flying objects for entertainment purposes. Ironically, the Mongolian army of Kublai Khan applied this same formulation for gunpowder and canons, which were used to destroy the Chinese strongholds and established the Yuan Dynasty (1271–1368).

One of the earliest mentions of the medical use of nitrate is in the treatment of what appears to be angina (chest pain) in an eighth century Chinese manuscript.[7] The patient was instructed to take niter by holding it under the tongue for a short period of time, and then swallowed the dissolved niter. According to Butler and Feelisch,[8] the significance of the instructions is that, under the tongue, nitrate-reducing bacteria convert some of the nitrate into nitrite.[9,10] It would then be the nitrite that dilates the blood vessels in the heart and alleviates the angina (see *nitric oxide and the two nitrate pathways*). For ancient Arabic healers and 12th century Salernian physicians alike, saltpeter was a regular component of their medicinal repertoire.[11] In the 17th and 18th centuries, nitrate attained such a status that it was included in various pharmacopeia. Its medical use included cure for ailments such as lack of appetite, kidney stones, prophylactic against small pox and even the bubonic plaque. In the 19th century, nitrate-based "traveler's powder" was sold in pharmacies and claimed to have curative power due to its anti-inflammatory and pain-killing property.

## Risks and Benefits of Nitrate

In terms of toxicity, nitrate is much less potent when compared to other natural occurring substances such as arsenic (Chapter 2), lead (Chapter 5), and mercury (Chapter 6). While nitrate was used as medicine throughout history, writings on its benefits and adverse effects were far fewer than that of arsenic, lead or mercury. Nevertheless, nitrate deserves our attention not only because it affects our health and well-being, but also because of the intense and passionate debate that is ongoing concerning the risks and benefits of nitrate and the appropriateness of setting guidelines values in food and water.

*Nitrate and methemoglobinemia.* In 1945, Hunter H, Comly[12] published an article entitled "Cyanosis in Infants caused by Nitrates in Well Water" which stated that cyanosis due to methemoglobinemia may occur when infants with gastrointestinal disturbance also received well-water with large amount of nitrate. This study sparked the interest of U.S. health agencies responsible for setting guideline values for nitrate. Since then other studies on nitrate have appeared with methhemoglobinemia remaining the primary focus. The maximum contamination level (MCL) of 44 mg/L was established by the US Environmental Protection Agency based on a survey by Walton[34] on nitrate concentration in water used for the preparation of feeding formula and incidences of methemoglobinemia. It was soon realized that nitrate and nitrite coexist in the environment as well as in the body, and if regulation is needed both should be addressed. The World Health Organization drinking water guidance level for nitrate is 50 mg/L, and 3 mg/L[14] for nitrite. Table 1 lists some of the current recommended guidelines for nitrate and nitrite in food and water.

**Table 1.** Guidance Limits on Nitrate and Nitrite in Water (mg Nitrate or Nitrite/L) and Food (mg Nitrate or Nitrite/kg Body Weight).

| | WHO 2011 | | JECFA (WHO) | | US EPA 2009 | |
|---|---|---|---|---|---|---|
| | Nitrate | Nitrite | Nitrate | Nitrite | Nitrate | Nitrite |
| Drinking Water | 50 | 3 | | | 44** | 3.3** |
| Food* | | | 3.7 | 0.07 | 7 | 0.33 |

*Reference dose (RfD) for US EPA and Acceptable Daily Intake (ADI) for WHO;
**Maximum Contamination Level (MCL).

*Nitrate and cancer.* In the body, nitrate is actively converted to nitrite, and *vice versa.* In acidic environments, such as the stomach, nitrite reacts readily with nitrostable compounds such as secondary amines and amides to generate N-nitroso compounds. Higher concentrations of N-nitroso compounds will of course be produced in the body if more nitrate, nitrite and nitrostable compounds are ingested. Some of the N-nitroso compounds formed in the body are known carcinogens. Whether nitrate and nitrite in food and water should be considered cancer causing substances has always been a hotly debated subject. A recent epidemiologic study conducted by the US National Cancer Institute[15] (Ward *et al.*, 2008) indicated that dietary nitrite and nitrate intake from animal sources were associated with significant increased risk for esophageal and stomach cancers. However, they did not observe a significant increase in risk of cancer in the population consuming public water containing higher nitrate and nitrite. A report by Andrew Mikowski and co-workers[16] at the University of Wisconsin and University of Texas represents a popular view on the subject. They reviewed up-to-date epidemiological data and concluded that evidence for cancer risk of nitrate, nitrite and processed meat are weak and inconclusive, and whatever risk is far outweighed by the health benefits of restoring nitric oxide homeostasis *via* dietary nitrite and nitrate. A review by the International Agency for Research on Cancer[17] concluded that ingested nitrate and nitrite *"under conditions that result in endogenous nitrosation"* is probably carcinogenic to humans (Group 2A). As far as drinking water is concerned, the World Health Organization conclusion is that nitrite itself is not carcinogenic to animals[18] and that for nitrate, *"...there is no convincing evidence of a causal association with any cancer site. The weight of evidence indicates that there is unlikely to be a causal association between gastric cancer and nitrate in drinking water"*.[14] As mentioned above the only guideline established by the World Health Organization for nitrate and nitrite is for drinking water, and they are based solely on methemoglobinia.

*Enterosalivary circulation of nitrate.* There is time at the dinner table when parents remind their children to chew their food thoroughly before swallowing. Older readers may recall a movement started by the American Horace Fletcher, who advanced the theory that chewing our food exactly 32 times adds strength to our body and reduces the amount of food required. He was considered a food faddist and nicknamed the "The Great Masticator".[18] Fletcher's advice turns out to have a grain of truth in it. Chewing is important because it breaks down the food into small pieces and aids in subsequent digestion. Food digestion actually starts in our mouth during chewing when salivary enzymes (amylase and lipase) initiate the process of breaking down carbohydrates and lipids, respectively. A third salivary enzyme, lysozyme, is a potent bactericide and helps to reduce the amount of microorganisms in the food ingested. In addition to these benefits of chewing we may now add a new one — chewing enhances the function of the enterosalivary circulation of nitrate. The core of the enterosalivary circulation is a group of symbiotic, nitrate reducing bacteria that reside on the dorsal surface of the tongue. These bacteria reduce nitrate present in the saliva and food to nitrite. Upon entering the acidic environment of the stomach, nitrite is converted to nitrogen oxide species which, as potent bactericides, destroy bacteria that enter the upper intestine. Nitrite and nitric oxides in the body are then converted back to nitrate and re-enter the salivary nitrate pool. Presence of nitrate in food and water are therefore beneficial to the host defense of the digestive system.

Presence of nitrate in food and water may also have other beneficial effects. For example, nitrate in circulation has been shown to modulate platelet activity in blood, and the motility and microcirculation of the gastrointestinal system.[19-21] G.M. McKnight and co-workers[19] propose the interesting hypothesis that *E. coli* O157, a strain often involved in outbreaks of foodborne illness, only emerged as a significant human pathogen after the introduction of regulation to limit the addition of nitrate to processed food.

***Nitric oxide and the two nitrate pathways.*** One of the most important discoveries of the past 30 years is the role of nitric oxide in the physiological functions of our body. The story begins in 1847 with the chemical synthesis of nitroglycerin, an organic nitrate, by Ascanio Sobrero of Turin, Italy. Sobrero noted the "violent headache" produced by minute quantities of nitroglycerin on the tongue.[22] In 1847, Alfred Nobel invented a safer and more reliable process for incorporating nitroglycerin into explosives and patented the product as "dynamite". According to W. Bruce Fye,[57] the first to experiment with nitroglycerin as a remedy was Constantine Hering, a German physician who moved to Philadelphia to practice and advocate homeopathic medicine. In 1849, Hering published his result on the investigation of nitroglycerin as a medicine for headache. The first clinical report on the ability of nitroglycerin to relieve angina pectoris was published by the British physician William Murell in 1879. A few years earlier Lauder Brunton had reported that an organic nitrite, amyl nitrite, was effective in reducing angina pectoris.[22]

For nearly a century, nitroglycerin was used extensively in the treatment of angina pectoris. However, other than its vasodilation effect, the mechanism of action of nitroglycerin was not known. In the 1970s, Ferid Murad investigated the action of vasodilators including nitroglycerin and concluded that nitric oxide released from nitroglycerin was the signal molecule that triggers vasodilation of blood vessels. At the same time Robert Furchgot was studying a labile compound named the endothelial-derived relaxing factor (EDRF) which controls blood vessel relaxation. Eventually, Louis Ignarro was able to prove experimentally that nitric oxide and EDRF were identical. This, of course, is a highly simplified description of the work originated from these three laboratories that ultimately revealed the key role of nitric oxide; and there are many scientists who contributed significantly to the field. In 1998, the Nobel Prize in Physiology and Medicine was awarded to Murad, Furchgott and Ignarro "*for their discoveries concerning nitric oxide as a signaling molecule in the cardiovascular system*". Alfred Nobel, if he were alive today, would have been gratified to know that the chemical that meant so much to him as an industrial dynamite would also be instrumental in ground-breaking research that was recognized by an award named after him.

In brief, what is known about the endothelial nitric oxide pathway is this: endothelial nitric oxide synthase in blood vessels (eNOS) generates nitric oxides from L-arginine. Nitric oxide diffuses (1) into the vascular smooth muscle cells and, through a series of molecular events, induces vasodilatation, and (2) into the blood stream where it inhibits inflammation and platelet aggregation. Nitric oxide is short-lived and is oxidized by tissue enzymes to nitrite and finally to nitrate.[23] The following scheme represents the endothelial nitric oxide pathway.

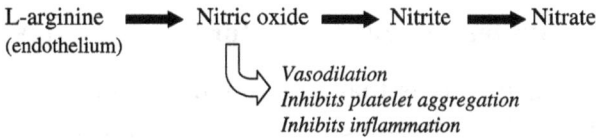

L-arginine ➡ Nitric oxide ➡ Nitrite ➡ Nitrate
(endothelium)

          ↳ *Vasodilation*
           *Inhibits platelet aggregation*
           *Inhibits inflammation*

In the above scheme, nitrate is considered an end metabolite serving no physiological functions. Recent research findings, however, suggested that we should not hold such a view anymore because there is evidence of a reverse pathway from nitrate to nitric oxide. First of all, the presence of nitrate-reducing bacteria on the tongue converts nitrate to nitrite in the mouth, which is converted to nitric oxides in the stomach (see *Enterosalivary circulation of nitrate*). Furthermore, recent evidence indicated that there are other *in vivo* reactions taking place in various tissues and organs that convert nitrite to nitric oxide.[24-26] The proposed reverse nitrate pathway is thus:

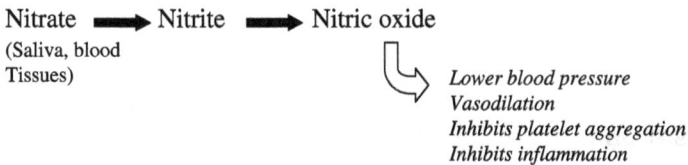

Nitrate ➡ Nitrite ➡ Nitric oxide
(Saliva, blood
Tissues)

          ↳ *Lower blood pressure*
           *Vasodilation*
           *Inhibits platelet aggregation*
           *Inhibits inflammation*

In this reverse pathway, nitrate is see as an active participant by serving as the substrate for the production of nitrite. Nitrite thus formed is considered a storage form of nitric oxide, and also a source of nitric oxide in the body that is independent of the endothelium enzyme mediated pathways. There is uncertainty about the physiological function of this non-endothelial nitric oxide. Recent studies suggest that it has beneficial function on blood vessels and lowers blood pressure in humans.[27–29] As we see in the following section, the presence of the reverse nitrate pathway forms the basis for criticisms of the regulatory limits set for nitrate in food and drinking water.

## Regulatory Guidelines and Limits for Nitrate and Nitrite

The major organizations that set guidance limits for nitrate and nitrite are the US Environmental Protection Agency (US EPA),[30] World Health Organizations (WHO)[17] and the Joint FAO/WHO expert Committee on Food Additives (JECFA).[31] The guidance values are listed in Table 1. The recommended daily limits for food intake, called reference dose (RfD) by the US EPA and allowable daily intakes (ADI) by WHO, are slightly different (Table 1). The RfD and the drinking water limits were all determined based on the estimated amounts of nitrate or nitrite that induce methemoglobinemia (see *Nitrate and methemoglobinemia*). The European Union (EU) also set limits for nitrate in vegetables and some baby foods. No one can fault the EU for not being zealous in their regulatory effort because the specifications not only specify the type of vegetable but also the growth season and whether cover was used during growth (Table 2). Lettuce and spinach grown under cover and in less sunny seasons tend to have higher nitrate content.[32,33]

## Criticism on Nitrate Limits

It is expected that the promulgation and setting of guideline limits for any substances will be met with resistance and disagreements. Criticism may come from scientists and medical professionals,

**Table 2.** Maximum Nitrate for Some Vegetables and Baby Foods (EFSA 2008).

| | | Maximum Level (mg nitrate/kg) |
|---|---|---|
| Spinach | Harvested October 1 — March 31 | 3,000 |
| | Harvested April 1 — September 30 | 2,000 |
| Preserved or frozen spinach | | 2,000 |
| Lettuce (*Latuca sativa. L*) | Harvested October 1 — March 31 | |
| | Grown under cover | 4,500 |
| | Grown in open air | 4,000 |
| | Harvested April 1 — September 30 | |
| | Grown under cover | 3,500 |
| | Grown in open air | 2,500 |
| Lettuce (Iceberg type) | Grown under cover | 2,500 |
| | Grown in open air | 2,000 |
| Processed cereal-based foods and baby foods | | 200 |

consumers, industry groups, or other stakeholders directly impacted by the guidelines and enforceable standards. Regulatory bodies view these criticisms as normal and constructive, and as valuable inputs in the revision processes. However, regulators setting guidelines for nitrite and nitrate must have been surprised by the number and intensity of response and criticism that followed.

The first published criticism probably comes from Jean L'Hirodel in the 1970s, whose studies and comments were collated and expanded in detail by his son Jean-Lois in the book *Nitrate in Man. Toxic, Harmless or Beneficial?*[11] The position of the book is clear on the outset as it stated in the preface "*But is it necessary to judge the quality of water by its content of nitrate according to an arbitrary figure that emerged one day from debates of an expert commission, referring to an impressive series of administrative texts closely copying each other, and thus giving the illusion of a consensus when everything is based on assumptions, not facts*". One can imagine the hard-working farmers, who had been relying for generations on wells with water containing naturally dissolved nitrate, being blind-sided by the regulations. Bias aside, the book presents careful constructed arguments and provides readers with a broad perspective of the subject of nitrate from different vantage points. L'Hirodel introduces us to the long history of nitrates in medicine. A chapter is devoted to the role of nitrate and the nitrogen cycle in nature and the necessity of nitrate for life. The next two chapters review nitrate metabolism and physiological roles of nitrate and nitrite in the body. The conclusion is that nitrate is a physiological important substance and need not be feared. The primary message of L'Hirodel was that regulatory limits on nitrate in food and water were, in their opinion, not supported by science and needed to be re-examined. He did not believe that the data produced by Comly and the American Association of Public Health, as summarized by Walton,[34] indicated a causal relationship between nitrate and infantile methemoglobinemia. He also took issue with the proposition that gastrointestinal disturbance causes increase in gastric pH, allowing accelerated conversion of nitrate by gastric bacteria and thereby increases nitrite production. The book's main conclusions may be summarized as follow: (1) Limits set on nitrate in drinking water was based on old and faulty epidemiological studies; (2) The disappearance of well-water methemoglobinemia in Western Europe and U.S.A. is likely due to the elimination of grossly unhygienic wells rather than to compliance with the nitrate standard; (3) Epidemiological studies do not support increased cancer risk with increased nitrate level; (4) WHO, EU and US standards on nitrate in water and food should be re-examined.

In 1999 Alexander Avery of the Center for Global Food Issues, Hudson Institute provided an insightful second look at the nitrate issue.[35] In agreement with L'Hironde, he suggests that nitrate in well-water is not the primary cause of methemoglobinemia; instead it serves as an indicator of bacterial contamination of the wells. Amongst the supporting evidence Avery quoted was a 1948 study by Marvin Cornblath and Alexis Hartmann[36] in which they fed water containing high levels (14 times present allowable daily intake) of nitrate to healthy infants and infants previously suffering from methemoglobinemia. After 6–9 days, only one of the four healthy infants showed elevated methemoglobin in their blood while the infants with previous history of methemoglobinemia had mild increase in methemoglobin and signs of moderate cyanosis. Such an experiment will of course be considered unethical today and would never have been approved for testing. Cornblath and Hartmann concluded from the result that there were other factors than ingested nitrate that determined whether or not an infant becomes cyanotic. Summing up studies available at the time, Avery concluded that infants with gastrointestinal infections produce high level of endogenous nitrate resulting in methemoglobinemia. The ingestion of large amount of nitrate exacerbates the condition but is not the primary cause. He warned that the setting of nitrate standards may give physicians a misconception that high nitrate in food and water is the main reason for methemoglobinemia, thereby ignoring other more important causes and treatments. In addition to the above health issues, Avery was also concerned with the costs incurred by municipalities to comply with an unnecessarily stringent nitrate standard. He proposed raising the nitrate standard for water from 44 mg/L to 66–89 mg/L.

McKnight and colleagues[19] challenged the premise that dietary nitrate is detrimental and therefore requires increasingly stringent regulations. They highlighted the antibacterial functions of the nitrate/nitrite enterosalivary circulation as a prime example of the beneficiary role of dietary nitrate.

Lorna Fewtrell's opinion article which appeared in a 2004 issue of the journal *Environmental Health Perspectives*[37] probably represented the first public divergence in opinion amongst experts who participated in the WHO drinking water guidelines. She investigated worldwide cases of drinking-water nitrate and methemoglobinemia and failed to find an exposure-response relationship. She observed that nitrate is only one of a number of co-factors that plays a complex role in the development of methemoglobinemia, and concluded that, given the low incidence of the disease, it is currently not appropriate to attempt to link illness rate with drinking-water nitrate levels. In other words, the premise based on which WHO and US EPA established their nitrate limits may not be very strong. In 2006, Deana Manassaram and co-workers[38] from the US Center for Disease Control and Prevention reviewed the literature on the effects of maternal exposure to drinking water nitrate. They concluded that current literature does not provide sufficient evidence of a causal relationship between nitrate and reproductive and developmental outcomes. In a study on pregnant women they did not find any association between methemoglobin levels and nitrate concentrations in drinking water.[39] This finding is consistent with that of Fewtrell *et al.*,[37] that there is not a well established quantitative exposure response relationship for nitrate exposure *via* drinking water and the development of methemoglobinemia.

In August 2004, a symposium[40] on drinking-water nitrate and health was held during the International Society for Environmental Epidemiology meeting in New York. The purpose of the meeting was to evaluate nitrate exposures and associated health effects in relation to the current regulatory limits. The possible involvement of nitrate in the *in vivo* generation of N-nitroso compounds, some of which are known carcinogens, was highlighted. It calls for more studies on the effects of nitrate on chronic diseases and cancer, especially on susceptible subgroups, and cautioned any premature decisions on changing nitrate limits. Similar call for more well-designed studies on exposure-response as well studies on the economic benefits of raising nitrate limits was issued by a group led by Hans van Grinsven of the Netherlands.[41]

The 2005 book *Nitrate, Agriculture and the Environment*[42] by T.M. Addiscott is a typical example of the reversal in opinion held by experts and opinion makers toward regulations on nitrate. Earlier in 1991, Addiscott published a book *Farming, Fertilizers and the Nitrate Problem*[4] in which they expressed the fear that "... *nitrate from agriculture had found its way into drinking water, causing stomach cancer in adults and cyanosis (methemoglobinemia) in infants ... a nitrate time bomb moving inexorably towards our taps.*" Now, 15 years later, Addiscott opined that much has changed because of the new understanding about nitrate.

In the Chapter on *Nitrate and Health,* Nigel Benjamin highlighted the beneficial role of what is now called the enterosalivary circulation for nitrate. Addiscott took issue with the "questionable" data used by WHO and US EPA to derive the nitrate limits and the scarcity of cases of infant methemoglobinemia. He questioned whether nitrate limits were justified. In the ensuing years, Addiscott and his colleagues[43,44] and others[45,46] continued to champion the cause that the potential health benefit of ingested nitrate outweighs the perceived hazards, and that the limits may be safely raised.

## New Research Findings

Investigators have been active in unearthing evidence of new health benefits of ingested nitrate. Jon Lundberg and his colleagues from Sweden have been one of the most active groups. They demonstrated in a 2006 study[47] that supplementation of healthy volunteers with sodium nitrate resulted in decreased diastolic but not systolic blood pressure. The finding was confirmed by Amrita Ahluwalia's group[48] in England who showed that human volunteers ingesting dietary nitrate (beetroot juice) had increased plasma nitrite concentration *via* bioconversion in the enterosalivary circulation. This bioactive nitrite substantially decreases blood pressure, inhibits platelet aggregation and prevents endothelial dysfunction. The blood pressure lowering effect was shown to be associated with an increase in nitric oxide converted *in vivo* from nitrite.[28,49] Furthermore, dietary nitrate has been shown to ameliorate pulmonary hypertension and that the effect was dependent on enzymes that convert nitrate to nitric oxide.[58] Recent work from Lundberg's group showed that dietary nitrate also reduces oxygen consumption during exercise which may be mediated through an effect on mitochondria.[50,51] In summary, the reported beneficial role of dietary nitrate on cardiovascular health in humans included blood pressure reduction, inhibition of platelet aggregation, vasoprotective activity, and reduced oxygen demand during exercise.[51] In addition, animal models showed that nitrite may protect the heart against ischemia-reperfusion damages.[52]

Not too long ago nitrate was viewed as simply a harmful compound.[4,53] Now, the beneficial effects are emerging and many believe that the benefits of nitrate outweighed its risk.[28,45,46,54,55]

## Response of Regulators to Concerns

Despite mounting criticisms, regulators and policy makers have been slow in their responses. The WHO guideline value for nitrate in drinking water is still 50 mg/L (WHO 2011).[14] That means the guideline values have not changed much since the landmark methemoglobinemia study of Comly in 1945.[12] On the other hand, there are signs that regulatory agencies are starting to re-evaluate their positions. Perhaps to address mounting dissatisfaction over limits imposed on vegetables, the European Food Safety Authority recently expressed the opinion that *Nitrate exposure at the current or proposed maximum levels for nitrate in spinach cooked from fresh is unlikely to be a health concern, although a risk for some infants eating more than one spinach meal in a day cannot be excluded.*[33] Although the nitrate limit was not revised, the World Health Organization[14] now recognizes that "...*there is compelling evidence that the risk of methemoglobinemia is primarily increased in the presence of simultaneous gastrointestinal infection...*" and that "*Nitrate probably has a role in protecting the gastrointestinal tract against a variety of gastrointestinal pathogens...It may have other beneficial physiological roles. There may therefore be a benefit from exogenous nitrate uptake, and there remains a need to balance the potential risks with potential benefits*".

Given the extent of criticism and the emergence of significant amount of new data, it is appropriate that there should be a review of the guidelines and limits for nitrate and nitrite. Most regulatory agencies have provisions for a revision or an update on their guidance documents. The WHO for example has a rolling revision provision for their guidance documents. The *Policies and Procedures Used in the Updating the WHO Guidelines for Drinking-Water Quality*[56] stipulates that in order to revise a guideline, it has to pass through 11 stages managed by six different groups (workgroups, offices, etc.) before approval. Any changes will be painstakingly slow and this should be true for most regulatory bodies.

## Practical Points

- If you depend on wells for drinking water, high nitrate levels (e.g. >50 mg/L) are always a concern. High nitrate may be a warning that the well is contaminated with other even more hazardous substances such as bacteria, fertilizers, pesticides.
- Methemoglobinemia in infants appears to be associated with high nitrate intake and concurrent diarrheal disease.
- Water should not be used for bottle-fed infants if the concentration of nitrate is above 100 mg/L but can be used if the concentration is between 50 and 100 mg/L if the water is microbiologically safe and there is increased vigilance by medical authorities.[14]
- Vegetables such as spinach, beet and lettuce may have high level of nitrate. Infants consuming food containing lettuce once a day is not harmful.[33]
- For adults, consumption of vegetables high in nitrate should not be curtailed because the health benefits of vegetables outweigh the potential hazards of nitrate and nitrite.
- Inappropriately stored vegetables can result in *in situ* conversion of nitrate to nitrite and should not be consumed.
- Nitrate is weakly associated with cancer of the stomach. Consumption of high amount of nitrate or nitrite treated/cured food is associated with an increase in cancer.
- Recent scientific studies indicate that nitrate have beneficial effects on health. There are appeals for regulatory bodies to revise the guideline values in drinking water and food.

## References

1. Canfield DE, Glazer AN, Falkowski PG. (2010) The evolution and future of earth's nitrogen cycle. *Science* **330**: 192–196.
2. Kasting JF, Siefert JL. (2001) Biogeochemistry: The nitrogen fix. *Nature* **412**: 26–27.
3. Spalding RF, Watts DG, Schepers JS *et al.* (2001) Controlling nitrate leaching in irrigated agriculture. *J. Environ. Qual.* **30**: 1184–1194.
4. Addiscott TM, Whitmore A, Powlson D. (1991) *Farming, Fertilizer and the Nitrate Problem.* CABI Publishing, Wallingford, UK.

5. Binkerd EF, Kolari OE. (1975) The history and use of nitrate and nitrite in the curing of meat. *Food Cosmet. Toxicol.* **13**: 655–661.

6. Lauer K. (1991) The history of nitrite in human nutrition: A contribution from German cookery books. *J. Clin. Epidemiol.* **44**: 261–264.

7. Cullen C, Lo V. (2005) *Medieval Chinese Medicine: The Dunhuang Medical Manuscripts.* Routledge Curzon, London, New York.

8. Butler AR, Feelisch M. (2008) Therapeutic uses of inorganic nitrite and nitrate: From the past to the future. *Circulation* **117**: 2151–2159.

9. Duncan C, Dougall H, Johnson P *et al.* (1995) Chemical generation of nitric oxide from the enterosalivary circulation of dietary nitrate. *Nat. Med.* **1**: 546–551.

10. Lundberg JO, Weitzberg E, Cole JA, Benjamin N *et al.* (2004) Nitrate, bacteria and human health. *Nat. Rev. Microbiol.* **2**: 593–602.

11. L'Hirondel J. and L'Hironel J-L. (2002) *Nitrate and Man: Toxic, Harmless or Beneficial?* CABI Publishing, New York.

12. Comly HH. (1945) Cyanosis in infants caused by nitrates in well water. *J. Am. Med. Assoc.* **129**: 112–116.

13. US EPA. (2009) National Primary Drinking Water Regulations. EPA 816-F-09-004. US Environmental Protection Agency; May 2009.

14. World Health Organization. (2011) *Guidelines for Drinking-Water Quality,* Fourth Edition. World Health Organization, Geneva.

15. Ward MH, Heineman EF, Markin RS *et al.* (2008) Adenocarcinoma of the stomach and esophagus and drinking water and dietary sources of nitrate and nitrite. *Int. J. Occup. Environ. Health* **14**: 193–197.

16. Milkowski A, Garg HK, Coughlin JR, Bryan NS. (2010) Nutritional epidemiology in the context of nitric oxide biology: A risk–benefit evaluation for dietary nitrite and nitrate. *Nitric Oxide* **22**: 110–119.

17. Background Document for Development of WHO Guidelines for Drinking-Water Quality. (2007) *Nitrate and Nitrite in Drinking-Water.* World Health Organization, Geneva. Report.

18. Christen AG, Christen JA. (1997) Horace Fletcher (1849–1919): The great masticator. *J. Hist. Dent.* **45**: 95–100.

19. McKnight GM, Duncan CW, Leifert C, Golden MH *et al.* (1999) Dietary nitrate in man: Friend or foe? *Br. J. Nutr.* **81**: 349–358.

20. Duncan C, Dougall H, Johnston P *et al.* (2005) Chemical generation of nitric oxide in the mouth from the enterosalivary circulation of dietary nitrate. *Nat. Med.* **1**: 546–551.

21. Govoni M, Jansson EA, Weitzberg E, Lundberg JO. (2008) The increase in plasma nitrite after a dietary nitrate load is markedly attenuated by an antibacterial mouthwash. *Nitric Oxide* 333–337.

22. Marsh N, Marsh A. (2000) A short history of nitroglycerine and nitric oxide in pharmacology and physiology. *Clin. Exp. Pharmacol. Physiol.* **27**: 313–319.

23. Nossaman VE, Nossaman BD, Kadowitz PJ. (2010) Nitrates and nitrites in the treatment of ischemic cardiac disease. *Cardiol. Rev.* **18**: 190–197.

24. Manukhina EB, Malyshev IY, Smirin BV *et al.* (1999) Production and storage of nitric oxide in adaptation to hypoxia. *Nitric Oxide* **3**: 393–401.

25. Cosby K, Partovi KS, Crawford JH *et al.* (2003) Nitrite reduction to nitric oxide by deoxyhemoglobin vasodilates the human circulation. *Nat. Med.* **9**: 1498–1505.

26. Li H, Cui H, Kundu TK *et al.* (2008) Nitric oxide production from nitrite occurs primarily in tissues not in the blood: critical role of xanthine oxidase and aldehyde oxidase. *J. Biol. Chem.* **283**: 17855–17863.

27. Lundberg JO, Gladwin MT, Ahluwalia A *et al.* (2009) Nitrate and nitrite in biology, nutrition and therapeutics. *Nat. Chem. Biol.* **5**: 865–869.

28. Kapil V, Milsom AB, Okorie M. (2010) Inorganic nitrate supplementation lowers blood pressure in humans: role for nitrite-derived NO. *Hypertension* **56**: 274–281.

29. Bahra M, Kapil V, Pearl V *et al.* (2012) Inorganic nitrate ingestion improves vascular compliance but does not alter flow-mediated dilatation in healthy volunteers. *Nitric Oxide.* In Press.

30. Nitrates and Nitrites. TEACH Chemical Summary. U.S. EPA, Toxicity and Exposure Assessment for Children's Health. [http://www.epa.gov/teach/chem_summ/Nitrates_summary.Pdf].

31. WHO Food Additives Series: 50. NITRATE and NITRITE. [http://www.inchem.org/documents/jecfa/jecmono/v50je07.htm].

32. European Food Safety Authority. Nitrate in Vegetables. (2008) Scientific opinion of the panel on contaminants in the food chain. *EFSA J.* **689**: 1–79.

33. European Food Safety Authority. (2010) Statement on possible public health risks for infants and young children from the presence of nitrates in leafy vegetables. *EFSA J.* **8**: 1935.

34. Walton G. (1951) Survey of literature relating to infant methemo-globinemia due to nitrate-contaminated water. *Am J Public Health Nations Health* **41**(8): 986–996.

35. Avery AA. (1999) Infantile methemoglobinemia: Reexamining the role of drinking water nitrates. *Environ Health Persp.* **107**: 583–586.

36. Cornblath M, Hartmann AF. (1948) Methemoglobinemia in your infants. *J. Pediatr.* **33**: 421–425.

37. Fewtrell L. Drinking-water nitrate, methemoglobinemia and global burden of disease: A discussion. *Environ. Health Persp.* **112**: 1371–1374.

38. Manassaram DM, Backer LC, Moll DM. (2006) A review of nitrates in drinking water: Maternal exposure and adverse reproductive and developmental outcomes. *Environ. Health Persp.* **114**: 320–327.

39. Manassaram DM, Backer LC, Messing R *et al.* (2010) Nitrates in drinking water and methemoglobin levels in pregnancy: A longitudinal study. *Environ. Health* **9**: 60.

40. Ward MH, deKok TM, Levallois P *et al.* (2005) International society for environmental epidemiology workgroup report: Drinking-water nitrate and health–recent findings and research needs. *Environ. Health Persp.* **113**: 1607–1614.

41. van Grinsven HJ, Ward MH, Benjamin N, de Kok TM *et al.* (2006) Does the evidence about health risks associated with nitrate ingestion warrant an increase of the nitrate standard for drinking water? *Environ. Health* **5**: 26.

42. Addiscott TM. (2005) *Nitrate, Agriculture and the Environment.* CABI Publishing, Wallingford, UK.

43. L'hirondel JL, Avery AA, Addiscott T. (2006) Dietary nitrate: Where is the risk? *Environ. Health Persp.* **114**: A458–A459.

44. Powlson DS, Addiscott TM, Benjamin N *et al.* (2008) When does nitrate become a risk for humans? *J. Environ. Qual.* **37**: 291–295.

45. Gilchrist M, Winyard PG, Benjamin N. (2010) Dietary nitrate–good or bad? *Nitric Oxide* **22**: 104–109

46. Machha A, Schechter AN. (2011) Dietary nitrite and nitrate: A review of potential mechanisms of cardiovascular benefits. *Eur. J. Nutr.* **50**: 293–303.

47. Larsen FJ, Ekblom B, Sahlin K *et al.* (2006) Effects of dietary nitrate on blood pressure in healthy volunteers. *N. Engl. J. Med.* **355**: 2792–2793.

48. Webb AJ, Patel N, Loukogeorgakis S *et al.* (2008) Acute blood pressure lowering, vasoprotective and antiplatelet properties of dietary nitrate via bioconversion to nitrite. *Hypertension* **51**: 784–790.

49. Jansson JA, Huang L, Malkey R. (2008) A mammalian functional nitrate reductase that regulates nitrite and nitric oxide homeostasis. *Nat. Chem. Biol.* **4**: 411–417.

50. Larsen FJ, Weitzberg E, Lundberg JO, Ekblom B *et al.* (2010) Dietary nitrate reduces maximal oxygen consumption while maintaining work performance in maximal exercise. *Free Radic. Biol. Med.* **48**: 342–347.

51. Larsen FJ, Schiffer TA, Borniquel S. (2011) Dietary inorganic nitrate improves mitochondrial efficiency in humans. *Cell Metab.* **13**: 149–159.

52. Webb A, Bond R, McLean P *et al.* (2004) Reduction of nitrite to nitric oxide during ischemia protects against myocardial ischemia-reperfusion damage. *Proc. Natl. Acad. Sci. USA* **101**: 13683–13688.

53. Tannenbaum SR, Correa P. (1985) Nitrate and gastric cancer risks. *Nature* **317**: 675–676.

54. Hord NG, Tang Y, Bryan NS. (2009) Food sources of nitrates and nitrites: The physiologic context for potential health benefits. *Am. J. Clin. Nutr.* **90**: 1–10.

55. Hord NG. (2011) Dietary nitrates, nitrites and cardiovascular disease. *Curr. Atheroscler. Rep.* **13**: 484–492.

56. WHO. (2009) WHO Guidelines for Drinking-water Quality. Policies and Procedures used in updating the WHO Guidelines for Drinking-water Quality. World Health Organization, Geneva.

57. Fye WB. (1986) Nitroglycerin: a homeopathic remedy. *Circulation* **73**: 21–29.

58. Baliga RS, Nilsom AB, Ghosh SM, *et al.* (2012) Dietary nitrate ameliorates pulmonary hypertension. Cytoprotective role for endothelial nitric oxide synthase and xanthine oxidoreductase. *Circulation* **125**: 2922–2932.

# CHAPTER 8

# CARBON DIOXIDE AND OXYGEN

The public normally do not consider oxygen and carbon dioxide to be health hazards. After all we live in an atmosphere consisting of a constant mixture of nitrogen, oxygen and carbon dioxide. We consume carbonated drinks regularly. Dry ice is freely used in stores and restaurants. Oxygen is regularly administered during emergency and medical procedures. Professional athletes invigorated themselves by inhaling oxygen. We breathe in oxygen, which is used by the body for aerobic respiration to generate energy, and exhale carbon dioxide. Plants, on the other hand convert carbon dioxide and water, in the presence of sunlight, into carbohydrates, and release of oxygen — a process called photosynthesis.

## Hazard of High Carbon Dioxide Levels

The air we breathe normally contains approximately 21% oxygen and 0.04% carbon dioxide by volume, while the air we exhale contains less oxygen (16%) and more carbon dioxide (4%). The adverse effects of acute exposure to carbon dioxide are summarized in Table 1. Volunteers inhaling air containing 2% carbon dioxide for an hour or more could experience headache and shortness of breath. As the concentration increases to 3.3 to 5%, increase in depth of breathing, abnormal electrocardiogram, and irritability are the common physiological responses. Even higher concentrations will produce adverse effects ranging from decreased mental performance and dizziness to abnormal muscle activity. When the carbon dioxide concentrations exceed 10%, breathing difficulty, loss of

131

consciousness and death may be the outcome depending on the duration of exposure.[1,2]

In the following sections, high carbon dioxide environments produced by natural disasters, and by man-made mistakes resulting in intoxication and death are described.

***Lake Nyos, Lake Monoun and Lake Kivu.*** The most publicized disaster involving carbon dioxide occurred in Lake Nyos, Cameroon.[3–5] Lake Nyos is a small lake several square km in area and more than 200 m deep. Its peaceful appearance belies the fact that it is a crater lake sitting on top of a now extinct volcano. Carbon dioxide-rich gas, which is continuously released through geothermal vents in the bottom of the lake, is trapped in the deep layers under high pressure. On 21 August 1986 between 9 and 10 p.m., a cataclysmic event took place — one side of a cliff surrounding the lake collapsed. The force of the slide was so strong that some of the debris reached the carbon dioxide-saturated layer of water at the bottom of the lake, causing disturbance in the deep layers and release of carbon dioxide gas. The technical term for the event is limnic eruption but most people call it a lake turnover. The eruption caused huge amount of carbon dioxide to be released into the atmosphere. Being heavier than air, the plume of carbon dioxide rolled down the valleys and engulfed four villages and the inhabitants. There was suggestion that the plume may have contained other toxic gases such as carbon monoxide. However subsequent analysis of dissolved gases at the bottom of the lake showed that the composition was >99.6% carbon dioxide with trace amounts of hydrogen sulfide, hydrogen cyanide, carbon monoxide and sulfur dioxide. The level of carbon dioxide at the instance when the toxic plumes engulfed the inhabitants will never be known, but it is likely near or exceed the 10% threshold when death may occur (Table 1). Of the affected inhabitants, 1746 died quickly from suffocation while 5,000 survived. Survivors showed symptoms of transient respiratory and lung effects. If there was any consolation, a follow up study conducted by a Cameroon medical team found that the survivors showed no lingering respiratory effects 55 months after the exposure.[5]

**Table 1.** Acute Health Effects of High Concentrations of Carbon Dioxide.[1]

| Carbon dioxide (%) | Duration | Effects |
| --- | --- | --- |
| 0.04 | | Normal atmospheric concentration |
| 1–1.15 | Days | Acid–base imbalance |
| 3.3 or 5.4 | 15 min | Increased depth of breathing |
| 6.0 | Several min | Abnormal electrocardiograms |
| 6.5 or 7.5 | 20 min | Decreased mental performance |
| 6.5 | 70 min | Irritability and discomfort |
| 7.5 | 15 min | Dyspnea, increased pulse rate, headache, dizziness, sweating, restlessness, disorientation, and visual distortion |
| 10 | 15 min | Eye flickering, excitation and increased muscle activity and twitching |
| >10 | Several min | Difficulty in breathing, impaired hearing, nausea, vomiting, a strangling sensation, sweating, stupor |
| >10 | 15 min | Loss of consciousness |
| 20–30 | Within min | Unconsciousness, convulsion, death. |

This was not the first time that such a tragedy occurred in Cameroon. A year earlier, on the early morning of August 15, 1985, another crate lake, Lake Monoun, turned over and the released plume killed 37 people.[4-7] Survivors' accounts pointed to a disturbance on the lake, which released a rolling white cloud as the initiating event. They described seeing people dying within minutes of being engulfed by the thick cloud. They recalled smells of rotten eggs[a] and gun powder. The most vivid account was from two survivors who were passengers on a pickup truck. As the truck drove

[a]Many years after the tragedy, experts were still debating whether carbon dioxide was the only culprit. The rotten egg smell suggested that hydrogen sulfide was present. Some expert suggested that carbon monoxide may also be involved. This debate may have been settled if blood samples were collected at the time from the victims because analysis of the haemoglobin can detect whether abnormal levels of hydrogen sulfide and carbon monoxide were present.

through the area the engine suddenly stopped. Nine people inside the truck got out to help the driver and all perished. Two persons who sat on top of the truck and did not come down survived. It is impossible to independently verify this oral testimony as more than a quarter of a century had transpired; nevertheless the account was consistent with carbon dioxide being the culprit. Indeed, scientists who sampled water for analysis after the eruption vividly described how the water samples pull up from the deep of the lake bubbled with gases. Carbon dioxide was heavier than air and settle on the road surface. It displaced the oxygen and hence choked off the truck engine. Two survived because they sat on top of the truck where the carbon dioxide level was lower.

To prevent the recurrence of these tragedies, teams of scientists and engineers had designed a procedure to degas Lake Nyos and Lake Monoun.[8,9] Pipes were lowered into the lake bottom. As the deepest layer of water was drawn to the surface, the pressure dropped and carbon dioxide was released. The mixture of degassed water and carbon dioxide bubbles was less dense and rose to the surface, drawing more water from the bottom and hence driving the degassing process continuously. The ultimate goal of this engineering undertaking is to reduce or eliminate future lake disruption.

For a period of time after the lake turnover, people evacuated from the disaster area. As time passed and memory of the tragedies faded, people started moving back to the vacated land. There is no way of assessing whether the degassing project is working, and it is impossible to predict when and if another lake eruption will take place.

In the meantime scientists have raised alarm on another lake — Lake Kivu, situated at the Eastern edge of the Democratic Republic of Congo. Similar to Lake Nyos, Lake Kivu trapped an enormous amount of carbon dioxide and methane at the bottom. Kivu is more than 3,000 times larger and contains more than 350 times as much gas as was released by Lake Nyos. There has been no record of a gas eruption (lake turnover) in Lake Kivu in recent history, and there is no consensus among scientists on the possibility of an occurrence. Nevertheless, a concern is justified because Lake Kivu is situated next to Mount Nyiragongo, an active volcano. The flow of lava from a 2002 volcanic eruption in fact reached the lake, although it did not sink deep enough to cause a lake turnover. If a bigger volcanic eruption does occur, two million inhabitants of Lake Kivu will face two potential catastrophes: carbon dioxide from an enormous lake turnover, and emissions and lava from the volcanic activity. In the mean-time, scientists and engineers are active around the lake. Regrettably, they are more interested in tapping the methane resource than preventing a carbon dioxide disaster.[10,11]

***Volcanic emission.*** Soil diffused degassing area surrounding volcanoes is a high-hazard zone because of the continuous emission of gases. The deadliest gas emitted is carbon dioxide because of the high concentration released. At a village near the Furnas volcano in the Azores Archipelagos, Portugal, indoor carbon dioxide as high as 21% had been measured at the floor level.[12]

In the past decades, hundreds of deaths occurred in volcanic areas which were directly attributable to $CO_2$. The most catastrophic event was the 1979 gas release at Dieng Plateau, Indonesia, which caused the death of at least 142 persons.[13] In the spring of 2011, the plateau showed signs of eruptive activity and the carbon dioxide level was starting to rise. The Indonesian government issued an alert and advised the inhabitants to vacate the area, but some chose to stay behind.[14] The seismic activity calmed down but the monitored carbon dioxide level remained high for some time. A remarkable photograph that appeared in the news media showed four children, probably on their way to or from school, with cloths wrapping around their nose and mouth. The cloths may ward off some volcanic dust but unfortunately will not stop gases such as carbon dioxide, hydrogen sulfide and carbon monoxide from entering the lungs.

Why did people move back to Lake Nyos and Lake Minoun soon after the lake turnover and why did some people refuse to evacuate the Dieng Plateau during volcanic activity? Much of the explanations must be sought in the social, economic and cultural domains. How people perceive risk and react in the face of natural hazards is a complicated field of study. We will have to leave the questions to experts in risk management and communication to solve.

***The Hidden Danger of Dry Ice.*** Dry ice is the solid form of carbon dioxide. It is common knowledge that because of the extremely low temperatures, it can cause skin burn if handled with bare hands, However there is a mistaken belief that because carbon dioxide is naturally present in the atmosphere, dry ice is harmless. In fact, exposure to dry ice in a confined or insufficiently ventilated space can be dangerous, even fatal. The following cases illustrate the hazards posed by dry ice in a variety of situations.

> Case 1. This accident occurred in Thailand. During a game of hide and seek, a child went inside a plastic container that had dry ice remnants. He was found in a convulsive state within five minutes and died soon after. The cause of death was attributed to carbon dioxide intoxication, and possibly in combination with asphyxiation due to lack of oxygen.[15]
>
> Case 2. A 59-year-old man entered a recently repaired walk-in freezer that was storing dry ice. He suffered a cardiac arrested and died. People who rushed in to help were unaware of the high carbon dioxide levels but fortunately were unaffected. The cause of death was determined to be carbon dioxide intoxication.[16]
>
> Case 3. As Hurricane Ivan approached Alabama in August 2004, a store owner decided to purchase dry ice for his perishable goods in anticipation of power outages. He picked up four blocks of dry ice totaling 100 lb and placed them in his pickup truck. The window of the vehicle was closed and the air conditioning was set to recirculate. After driving for half a mile the man suffered shortness of breath, which became worse as he drove on. He telephoned his wife to call the emergency number, pulled up to a parking lot and promptly lost consciousness. His wife located him in the parking lot soon after. The man regained consciousness when his wife opened the door of the vehicle. He recovered completely and complained only of a headache.[17]

There are additional stories involving people working in special effect units,[18] workplace freezers,[19] and other occupational situations.

Although manufacturers issued warning about the hazards and instructions on how to handle and store dry ice, the general public appeared to be unaware of the danger.

## Hazard of Low Oxygen Levels

Table 2 summarizes the adverse effects of low oxygen levels on our body. When the oxygen concentration in air is reduced to 12–16% by volume, the physiological response is to increase the breathing rate and the pulse rate. When oxygen levels decreased to 10–14%, more serious effects such as fatigue, mental disturbance and respiratory abnormality occurred. At even lower oxygen levels, loss of consciousness and death could be the final outcome. Injuries and death due to asphyxiation are usually associated with people working in confined or poorly ventilated spaces. There are also many anecdotal cases of children suffering from asphyxiation because of accidental entry into containers and refrigerators.

As the altitude increases, the air is thinner and has less oxygen. Table 3 describes the relationship between latitude, oxygen concentration, and various altitude diseases. To climbers of Mount Everest, the altitude above 8,000 m (26,200 ft), where the oxygen level does not support life, is known as the death zone (Table 3). Paul Firth and co-workers[36] reviewed the causes of mortality on Mount Everest during the years 1921–2006 and noted that the majority of climbers died during descent from the summit. The victims typically

**Table 2.**   Adverse Effects of Short-term Oxygen Deficiency.[1]

| Oxygen level (by volume) | Effects |
|---|---|
| 21% | Normal |
| 12–16% | Breathing and pulse rate are increased, slight muscular in-coordination |
| 10–14% | Emotional upsets, abnormal fatigue from exertion, disturbed respiration |
| 6–10% | Nausea and vomiting, inability to move freely, collapse, possible lack of consciousness |

**Table 3.**　Altitude, Oxygen Level and Acute Effects on Health.

| Altitude (ft) | Oxygen in air (%) | Landmarks | Health Effects |
|---|---|---|---|
| 0 | 21 | Sea level | None |
| 5000 | 18 | Tibet (low plain), Denver city | Some may experience discomfort |
| 8000 | 16 | Macu Picchu | Acute Mountain Sickness, High |
| 16,000 | 12 | Tibet (High plain), Mt. Blanc | Altitude Pulmonary Edema, High Altitude Cerebral Edema |
| 20,000 | 10 | Mt. Kilimanjaro, Mt. McKinley | |
| 29,000 | 8 | Mt. Everest | Does not support life |

suffered cognitive impairment and ataxia, symptoms of high altitude cerebral edema. They postulated that the death was caused by exhaustion of supplemental oxygen supplies during descents. In May, 2012, 4 climbers from Germany, South Korea, China and Canada died during descents from the summit of Mount Everest. Spring was a particular busy time and climbers had to wait a long time for their turn to ascent. On their way down, the supplemental oxygen supplies were exhausted, and acute hypoxia set in. The following sections describe the history and characteristics of acute and chronic mountain sicknesses.

***Acute Mountain Sickness.*** According to Daniel Gilbert of the U.S. National Institute of Health, the first historical account of mountain sickness was recorded by Too Kin, an official in the West Han Dynasty of China.[20] In his description of an expedition (37–32 BC) to what is now the Pamir Plateau of Xinjiang he wrote: "*Next, one comes to Big Headache and Little Headache Mountains, as well as Red Earth and Swelter Hills. They make a man so hot that his face turns pale, his head aches, and he begins to vomit. Even the donkeys and swine react this way*".[20] While the exact location of the mountains cannot be identified, Gilbert proposed several possible routes the expedition force may have taken and one of these routes reached an altitude as high as 16,000 ft. Members of the expedition were probably conscripts from low-lying areas of China and had no idea that they were afflicted with a disease called acute mountain sickness (AMS) (Table 3).

A more detail account of AMS came from the Spanish Jesuit Joseph de Acosta (1539–1599) who went to Peru in 1599. Samuel Jarcho[21] relayed Acosta's experience thus: "... *when I went up what are called "the stairs" — the highest part of the range almost immediately I suffered anguish so severe that I had the idea of throwing myself from the cavalcade on to the ground ... At the same time I retched and vomited so violently that I thought I was dying, because after throwing up food and phlegm, then bile and more bile, some yellow and some green, I finally vomited blood, on account of the violence which my stomach was suffering. If this had lasted I should certainly have perished, but it did not continue for more than three or four hours, until we had descended a considerable distance and arrived at a more suitable climate ...I am convinced that the element of air in the region of Pariacaca is so subtle and delicate that it is unsuitable for human respiration, which requires thicker and more tempered air*". Using modern maps, Gilbert[22] identified the route in the Andean mountain range traversed by Joseph de Acosta; The maximum altitude of his trail reached 4,800 m (15,750 ft), about the same elevation as the summit of Mt. Blanc, the highest mountain in Western Europe (Table 3).

Today we know more about altitude-related disease. AMS typically occurred at altitude above 8,000–9,000 ft when the partial pressure of oxygen is about 75% of sea level or less (Table 3). Within 6–12 hours of arriving at that altitude, some travelers will experience headache, intestinal upset, fatigue, and difficulty sleeping. The symptoms usually resolve on their own within 2–3 days. At higher altitudes, AMS may progress to more severe complications called high altitude pulmonary edema (built up of fluid in the lungs) and high altitude cerebral edema (swelling of the brain).[23-25] Both of these complications are life-threatening.

Scientist at the International Hypoxia Symposium have adopted a Lake Louis AMS Questionnaire for the diagnosis of AMS.[25] A diagnosis is established if the following conditions are met (1) ascent to high altitude in the past four days; (2) onset of headache; and at least one of these symptom; gastrointestinal disturbances, fatigue and or weakness, dizziness/light-headedness, difficulty sleeping.

With increasing interest in new and exotic vacation spots, more and more people are travelling to high altitude locations. In the past, it took days or even weeks to arrive at a high-altitude destination by motor vehicles or by foot. Today, it takes just a few hours for travelers to reach a high mountain or plateau by air. They simply do not have the time to adjust to the thin air. A 2010 survey was conducted on risk of illness and injury occurring in members participating in commercial expeditions.[26] It was found that more than half (51%) of the serious medical incidents documented was due to AMS with signs of high altitude cerebral/pulmonary edema.

The best practice to prevent diseases associated with high altitude is to acclimatize by ascending gradually. A rule of thumb is to ascend 1,000 ft per day for altitude above 8,000–10,000 ft. For those travelling by plane, they should avoid physical exertion in the first few days to allow the body to acclimatize to the low oxygen levels. Individuals with one or more of the diseases associated with high altitude should descend immediately. Persons with preexisting medical conditions such as cardiovascular, respiratory and renal and diseases, diabetes, anemia and psychiatric disorders should consult a physician before travelling to high altitude destinations.[27]

***Chronic Mountain Sickness (CMS).*** Much attention are paid to AMS because it affects individual explorer and traveler to high altitude destinations and because the effects are immediate and drastic. However, in terms of the total number of people affected, chronic mountain sickness (CMS) should receive as much attention if not more.[28] The disease was also called the Monge's disease in honor of Carlos Monge, a Peruvian physician (1884–1970) who pioneered the research in high altitude medicine. It was estimated that 140 million people lived in high altitude regions (>8,200 ft). Of this, 80 million are in Asia and 35 million are in the Andean mountains.[29] After prolonged stay at such high altitude, 5–15% of the population ultimately developed CMS. The sickness is characterized by polycythemia (increased red blood cells), hypoxemia (low oxygen tension in the blood), and pulmonary hypertension. Common manifestations of CMS are headache, dizziness, tinititus, breathlessness, fatigue, anorexia, mental confusion, cyanosis, and dilation of veins.

The most effective treatment of CMS is to have the patients move to a lower altitude where their sickness gradually resolves. Carlos Monge and others suggested that coca leaves chewed by natives in the high Andes may help prevent the onset of CMS, but this claim was never confirmed by medical studies. Studies have suggested that azetazolamide alone or in combination with dexamethasone may be effective in treating CMS, but more verification studies are needed.[30]

## Hazard of High Oxygen Levels

We are not exposed to excessively high levels of oxygen because the level in the environment is quite constant. However, there are therapeutic procedures which use high levels of oxygen, sometime under high pressure and oxygen-related toxicity may occur. Premature infants who require hyperoxic respiratory support sometimes develop acute and chronic eye and lung changes due to oxygen toxicity.[31,32] Adults on oxygen therapy for a prolonged period of time may develop complications that include visual impairment, altered respiratory physiology, and retinal and pulmonary injuries.[31,33] Hyperbaric oxygen therapy is routinely used to treated patients with carbon monoxide poisoning. One of the complication of the treatment is cerebral seizure which is attributed to the toxicity of oxygen on the central nervous system.[34,35]

## Practical Points

- Dry ice is an often ignored source of carbon dioxide poisoning. Some safety tips:
  — Use and store dry ice in well-ventilated area.
  — Do not transport dry ice in vehicle with windows closed.
  — Make sure containers with dry ice are not accessible to children.

- Air masks do not prevent inhalation of carbon dioxide.
- Avoid working in confined space where ventilation is inadequate. The hazards are high carbon dioxide level and depletion of oxygen supply.
- Individuals with pre-existing medical conditions should consult a physician before travelling to high altitude destinations.
- To avoid contracting AMS, avoid ascending to high altitude too quickly.
- Individuals who contracted AMS should move to lower altitude as soon as possible.

## References

1. Canadian Centre for Occupational Health and Safety (CCOSH). [http://www.ccohs.ca/oshanswers/chemicals/chem_profiles/].
2. Langford NJ. (2005) Carbon dioxide poisoning. *Toxicol. Rev.* **24**: 229–235.
3. Kling GW, Clark MA, Wagner GN *et al.* (1987) The 1986 Lake Nyos gas disaster in Cameroon, West Africa. *Science.* **236**: 169–175.
4. Sigurdsson H. (1988) Gas bursts from Cameroon crater lakes: A new natural hazard. *Disasters* 131–46.
5. Wagner GN, Clark MA, Koenigsberg EJ, Decata SJ *et al.* (1988) Medical evaluation of the victims of the 1986 Lake Nyos disaster. *J. Forensic Sci.* **33**: 899–909.
6. Kling GW. (1987) Seasonal mixing and catastrophic degassing in tropical lakes, Cameroon, West Africa. *Science* **237**: 1022–1024.
7. Killer Lakes. (2002) *British Broadcasting Corporation, Science and Nature.* 9:00 pm, Thursday 4, August [http://www.bbc.co.uk/science/horizon/2001/killerlakes.shtml].
8. Clarke T. (2001) Hazard management. Taming Africa's killer lake. *Nature* **409**: 554–555.
9. Kling GW, Evans WC, Tanyileke G *et al.* (2005) Degassing lakes Nyos and Monoun: Defusing certain disaster. *Proc. Natl. Acad. Sci. USA* **102**: 14185–14190.
10. Sigurdsson H. (1988) Gas bursts from Cameroon crater lakes: A new natural hazard. *Disasters* **12**: 131–146.

11. Najar A. (2009) A lakeful of trouble. *Nature* **460**: 321–323.

12. Viveiros F, Ferreira T, Silva C, Gaspar J *et al.* (2009) Meteorological factors controlling soil gases and indoor $CO_2$ concentration: A permanent risk in degassing areas. *Sci. Total Environ.* **407**: 1362–1372.

13. Le Guerin F, Tariff H, Favre Pierre R. (1982) An example of health hazard: People killed by gas during a prelatic eruption: Ding plateau (Java, Indonesia), February 20th 1979. *Bull. Volcano.* **45**: 153–156.

14. People Near Timbang Crater Won't Leave Despite Gas Threats. (2011) *The Jakarta Post.*

15. Srisont S, Chirachariyavej T, Peonim AV. (2009) A carbon dioxide fatality from dry ice. *J. Forensic Sci.* **54**: 961–962.

16. Dunford JV, Lucas J, Vent N *et al.* (2009) Asphyxiation due to dry ice in a walk-in freezer. *J. Emerg. Med.* **36**: 353–356.

17. Brief Report: Acute Illness from Dry Ice Exposure during Hurricane Ivan—Alabama, 2004. (2004) *MMWR.*

18. Hsieh CC, Shih CL, Fang CC *et al.* (2005) Carbon dioxide asphyxiation caused by special-effect dry ice in an election campaign. *Am. J. Emerg. Med.* **23**: 567–568.

19. Gill JR, Ely SF, Hua Z. (2002) Environmental gas displacement: Three accidental deaths in the workplace. *Am. J. Forensic Med. Pathol.* **23**: 26–30.

20. Gilbert DL. (1983) The first documented report of mountain sickness: The China or Headache Mountain story. *Respir. Physiol.* **52**(3): 315–26.

21. Jarcho S. (1958) Mountain sickness as described by Fray Joseph de Acosta, 1589. *Am. J. Cardiol.* **2**: 246–247.

22. Gilbert DL. (1983) The first documented description of mountain sickness: The Andean or Pariacaca story. *Respir. Physiol.* **52**: 327–347.

23. Bia FJ. (1992) Current prevention and management of acute mountain sickness. *Yale J. Biol. Med.* **65**: 337–341.

24. Wilson MH, Newman S, Imray CH. (2009) The cerebral effects of ascent to high altitudes. *Lancet Neurol.* 175–191.

25. Maa EH. (2010) Hypobaric hypoxic cerebral insults: The neurological consequences of going higher. *NeuroRehabilitation* **26**: 73–84.

26. Lyon RM, Wiggins CM. (2010) Expedition medicine–the risk of illness and injury. *Wilderness Environ. Med.* **21**: 318–324.

27. Mieske K, Flaherty G, O'Brien T. (2010) Journeys to high altitude–risks and recommendations for travelers with preexisting medical conditions. *J. Travel Med.* **17**: 48–62.

28. León-Velarde F. (2003) Pursuing international recognition of chronic mountain sickness. *High Alt. Med. Biol.* **4**: 256–259.

29. Penaloza D, Arias-Stella J. (2007) The heart and pulmonary circulation at high altitudes: Healthy highlanders and chronic mountain sickness. *Circulation* **115**: 1132–1146.

30. Bernhard WN, Schalick LM, Delaney PA *et al.* (1998) Acetazolamide plus low-dose dexamethasone is better than acetazolamide alone to ameliorate symptoms of acute mountain sickness. *Aviat. Space Environ. Med.* **69**: 883–886.

31. Winter PM, Smith G. (1972) The toxicity of oxygen. *Anesthesiology* 210–241.

32. Frank L. (1985) Effects of oxygen on the newborn. *Fed. Proc.* **44**: 2328–2334.

33. Smith G, Shields TG. (1975) Oxygen toxicity. *Pharmacol. Ther. B* **1**: 731–756.

34. Bitterman N. (2004) CNS oxygen toxicity. *Undersea Hyperb. Med.* **31**: 63–72.

35. Stoller KP. (2007) Hyperbaric oxygen and carbon monoxide poisoning: A critical review. *Neurol. Res.* **29**: 146–155.

36. Firth PG, Zheng H, Windsor JS, *et al.* (2008) Mortality on Mount Everest 1921–2006: a descriptive study. *Br. Med. J.* **337**: a2654.

# CHAPTER 9

# TOXIC GASES OF NATURE

## Sulfur Dioxide

### Sources and health effects

Sulfur dioxide ($SO_2$) is a colorless gas with a pungent odor. Before the industrial revolution, natural emissions were likely the primary source of atmospheric $SO_2$. Today, the largest sources of emissions come from fossil fuel combustion at power plants (73%) and other industrial facilities (20%). Smaller sources of $SO_2$ emissions include industrial processes such as extraction of metal from ore, and the burning of sulfur-containing fuels by locomotives, ships, and other non-road devices. In industrialized countries such as the U.S., the 24-hour average $SO_2$ concentration is typically less than 0.1 ppm, with short-term levels reaching more than 1 ppm.[1-3]

Because of its high solubility, inhaled $SO_2$ is readily absorbed through the upper respiratory tract where it causes rapid bronchoconstriction. While the exact mechanism of $SO_2$-induced bronchoconstriction is not completely understood, there is evidence to suggest that both parasympathetic pathways and inflammatory processes are involved, which are potentiated among asthmatics through the mediation of sensory receptors and mast cells in the airways.[87,88] Exposure of between 1 and 6 hours to concentrations as low as 1 ppm has been shown to produce a reversible decrease in lung functions. Exposure to elevated concentration of $SO_2$ for less than 24 hours can cause reduction in forced expiratory volume and other indices of ventilatory capacity, increased airway resistance, and symptoms such as wheezing and shortness of breath. Studies show a connection between short-term exposure and increased visits to

145

emergency departments and hospital admissions for respiratory illnesses, particularly in at-risk populations such as children, the elderly, and asthmatics.[2] Besides asthmatics, children and adults performing outdoor physical activities for prolonged period of time are also vulnerable to the respiratory effect of $SO_2$.

The air quality standards for $SO_2$ set by the World Health Organization are: 0.0075 ppm (20 $\mu g/m^3$) for 24-hour exposure and 0.019 ppm (500 $\mu g/m^3$) for 10-min exposure.[3] Some countries have maintained their own standards, e.g. the U.S. standard is 0.075 ppm for 1-hour exposure. The limit of 0.14 ppm for 24-hour and 0.03 ppm for annual exposure were revoked in 2010.[1] In addition, the U.S. National Park Service has set its own criteria for issuing air quality advisories: Good, <0.2 ppm; moderate, ≥0.2 ppm; unhealthy for sensitive people, ≥0.4 ppm; unhealthy, ≥1ppm; very unhealthy ≥3 ppm; hazardous ≥5 ppm.[5]

The major natural sources of atmospheric $SO_2$ are volcanic gases, geothermal emissions and wild fire emissions. While the relative proportion varies with the geographical location of eruption, $SO_2$ is generally the third most abundant volcanic vapors emitted after water and carbon dioxide.[4] Ambient level as high as 2.4 ppm has been registered at a Hawaii Volcanoes National Park monitoring site.[5]

### Stories and scientific accounts

**Smoking Hills.** In search for a northwest passage through the Arctic Ocean in 1826, the English explorer Sir John Franklin observed clouds of white smoke, and rocks and soil that appeared to be on fire on the cliffs of what is now known as Cape Bathurst, Northwest Territories. Legend has it that Franklin named the place the Smoking Hill. Decades later, Captain Robert McClure, on a mission to find the missing Sir Franklin and his crew, returned to the same location. Sailors were said to have returned from the shore with samples of the smoldering rock, which promptly burned a hole on Captain McClure's desk. For centuries, masses of carbon-rich shales and pyrite (iron sulfite) along 30 km of this arctic coastline have been literally igniting spontaneously as the hills erode and mineral veins are exposed to air, producing plumes rich in $SO_2$ and acidic aerosols. For miles around, the soils are severely acidified by $SO_2$ and therefore devoid of vegetation and animals. Ponds in the surrounding area have pH measurements averaging 2–3 instead of the normal pH of 8. Health impact on humans is minimal because the area is sparsely populated; the nearest settlement is more than 100 km away. The strikingly lifeless expanse nevertheless serves to illustrate what excessive $SO_2$ from nature can do to the environment and all living organisms.[6,7,88]

***Emissions of $SO_2$ from Volcanic and Geothermal Eruption.*** On April 14, 2010, eruptions of the Eyjafjallajökull volcano in Iceland sent fine-grained ash into the jet stream. The ash quickly dispersed over Europe and disrupted air traffic for weeks. The Eyjafjallajökull event serves as a reminder that volcanic activity can affect our activity, health and well-being at any given time. It is estimated that at least 455 million people worldwide live within potential exposure range of a volcano that has been active within recorded history.[8]

One of the largest volcanic eruptions occurred in Iceland in the summer of 1783, when fissures opened on each side of the Laki crater. During an eight-month period an estimated $122 \times 10^{12}$ g of $SO_2$ and smaller amount of other gaseous components were released to the atmosphere from a series of explosive fissures and vents, and from the voluminous lava flows. The eruption caused environmental stress and posed serious health hazards far beyond the borders of Iceland.[9,10] Historical accounts of increased respiratory disorders and mortality were found in the Netherland, France, Italy and Sweden.[11] In England alone, analysis of parish records indicated that there were two distinct mortality crisis periods: August–September 1783, two months after the beginning of the eruptions, and January–February 1784, which together accounted for ~19,700 extra deaths. It has been suggested that the fine acid aerosol and gases from the volcanic haze, which was also commonly called the "great dry fog", was partly responsible for the excess mortality.[12]

While the suggested link between $SO_2$ emission and mortality in the case of the Laki eruption was hard to prove because of the obvious absence of chemical data, studies on more recent and smaller scale volcanic eruptions appeared to support a correlation with health effects. Following a volcanic eruption in 2000, the ambient $SO_2$ on the island of Miyakejima remained elevated for many years. A study conducted on healthy volunteers who returned to the island for six months in 2005 and were exposed to a range of $SO_2$ concentration between 0 and 10 ppm found a statistical significant exposure-response relationship between $SO_2$ and acute respiratory symptoms, with females and non-smokers showing higher sensitivity to the irritant gas.[13] Another study focusing on adults two years after returning to the island and exposed to average $SO_2$ concentration between 0.019 to 0.045 ppm observed minor effects on the respiratory system.[14] The Kilauea Volcano on the island of Hawaii has vented continuously since 1983 but on March 2008 a new and powerful eruption occurred, raising the average daily $SO_2$ concentration to as high as 0.075 ppm, 10 times the 24-hr limit of 0.0075 ppm recommended by the World Health Organization.[3] Subsequent study observed a statistically significant increase in hospital visits for medically diagnosed coughs, headaches, acute pharyngitis, and acute airway problems. There was also evidence of increased morbidity of acute illnesses. Approximately three months after the major eruption, average 24-hour outdoor $SO_2$ concentrations still remained elevated.[15,16,83,84]

### Practical points

- Pay attention to news bulletins and air pollution advisories concerning $SO_2$.
- Limit outdoor activities during times of high air pollution.
- If you live in an area affected by volcanic and geothermal emissions, pay attention to the advice of health and public officials about the air pollution situation.
- Children, asthmatics, elderly individuals, and people with respiratory diseases (e.g. asthma), chronic obstructive pulmonary diseases and cardiovascular diseases are more susceptible to adverse effects of $SO_2$ exposure.

- People performing heavy physical activity or exercise outdoor and children spending a lot of times outdoors are more vulnerable to the effects of $SO_2$ exposure.

## Hydrogen Sulfide

### Sources and historical accounts

Hydrogen sulfide ($H_2S$) is a colorless, flammable gas. It has a characteristic rotten egg odor that is detectable at very low concentration (Table 1). With a vapor density of 1.19, $H_2S$ is approximately 20% heavier than air and can collect in low-lying areas and in confined spaces. $H_2S$ is produced from industrial activities such as food processing, paper mills, coke oven plants, tanneries, sulfuric acid and inorganic sulfide manufacturers, and petroleum refineries. $H_2S$ occurs naturally in crude petroleum, natural gas, volcanic gases, geothermal vents, undersea vents and sulfur springs. Hydrogen

**Table 1.** Effects of $H_2S$ at Various Concentration.[18,32,82]

| Concentration (ppm) | Signs & Symptoms |
| --- | --- |
| 0.01–0.3 | Odor threshold (highly variable) |
| 2 | Breathing problem in asthmatics |
| 1–5 | Increased eye complaints; may be associated with headache, nausea and loss of sleep with prolonged exposure |
| 5–10 | Problems with energy metabolism |
| 4–20 | Eye irritation, coughing, throat irritation |
| 20 | Fatigue, dizziness, headache, poor memory, irritability, loss of appetite |
| >100 | Loss of sense of smell, severe eye and lung irritation |
| >250 | Respiratory distress, lung damage |
| >500 | "Knockdown" (sudden unconsciousness) and death (within 4–8 hr) |
| 1000 | Breathing may stop with one or two breaths, immediate collapse, death |

sulfide is commonly found in coal, natural gas and petroleum deposits and may be mobilized by human activities such as mining and drilling. Natural gas containing significant amount of $H_2S$ is called sour gas. Microbial breakdown of human and animal waste with sulfur and sulfate-containing organic compounds produces $H_2S$. It may be present in high concentrations in sewer gas associated with sewer system and treatment plants, as well as in manure gas associated with pig farming. Landfill gas may also contain $H_2S$. Swamp gas or marsh gas resulting from bacterial breakdown of organic matters also contain varying concentration of $H_2S$.[17,18]

***Ramazinni's account of sewer gas.*** The first documented account of the effects of sewer gas was written by the Italian physician, Bernardino Ramazzini (1633–1714), who was recognized as the father of occupational medicine. He noted that the eyes of sewer workers were extremely bloodshot and dim, and made the following astute observation "*I am inclined to think that some volatile acid is given off by this camerine of filth when they disturb it, and what makes this probable is the fact that copper and silver coins carried by these scavengers in their purses turn black*".[19] Ramazzini would not have known at the time but he had put his finger on the gas that caused the sewer worker's eye problems. Moreover his observation on the tarnished coins is absolutely correct. Indeed the reaction of $H_2S$ with silver produces the black-colored silver sulfide — a reaction that was utilized by scientists later on to assay for silver and $H_2S$.

***Birds and toxic gases.*** A swamp is an unfriendly place. It is inhabited by dangerous animals and organisms. Medical literature is replete with information about the risk of contracting various tropical or infectious diseases from mosquitoes or water-borne parasites. We might also have heard tales of the toxic and potentially deadly swamp gas. However, very little information can be found in the scientific and historical literature regarding the presence and toxicity of swamp gas. A Chinese historical account of the adventure of General Ma Hwan more than two thousand years ago comes close to a description of swamp gas. At the beginning of the first century AD, General Ma was sent by the Emperor of the Han dynasty to lead an army to suppress a rebellious vassal state in southern China. The first expedition was a failure because, in addition to the enemy, the troops had to overcome treacherous terrain, warm and humid climate and toxic "vapor" — *Alas, the Wu River* — *poisonous and insidious!* wrote General Ma. Learning from his earlier mistake, the second campaign was a success. During his victory celebration Ma wrote this poem:

> *Mist above the watery quagmire*
> *Toxic fumes steamy and dense*
> *Look up and behold,*
> *flying birds falling unconscious from the sky*

Flying birds falling unconscious from the sky? Did General Ma exercise his poetic license too freely, or is there a connection between toxic gas and birds falling from the sky? Based on what we know today about the nature of swamps and the sensitivity of birds to toxic fumes, the answer is probably yes. First, in the steamy heat of the tropic and sub-tropic, dead vegetations and animal matters in swamps undergo a natural decaying process aided by bacteria and other microbes, releasing a number of gases with the most prominent being $H_2S$. The *toxic fume steamy and dense* refers to the swamp gas; $H_2S$ is more dense than air and stays low — a mist hugging the swamp surface. Second, birds are highly sensitive to toxic fumes. For example, canaries had been routinely used in coal mines to detect carbon monoxide.[20]

Legend has it that it was at the Staffordshire collier on 27 May 1901, that rescuers first brought the canaries along with them down the mine shafts. The era came to a close in 1986 when 200 canaries were retired from British mining pits, being replaced by carbon monoxide detectors (BBC).[21] Stories about pet birds succumbing to fumes emitted from Teflon frying pans left too long on the stove are true, not an urban myth. Scientific literature were full of reports describing avian species from pet birds to chickens killed by fumes generated from overheated utensils coated with polytetrafluoroethylene (PTFE) polymers.[22–24]

The real question is: can birds be used as sentinels for toxic gases? We know they are sensitive to carbon monoxide and PTFE fumes, but are they more sensitive than human to $H_2S$ and other toxic gases? Today's media is filled with stories of "birds falling from the sky". While some of the phenomena have been explained, many incidences remain involved and significantly more research is needed in this area.

***Suicide by Hydrogen Sulfide.*** Hydrogen sulfide is easily generated by mixing liquid bath essences containing sulfur with household cleaners containing hydrochloric acid. Sadly, this method has been used by people to commit suicide. Japan is the country with the most documented cases of suicide by $H_2S$ partly because of the popularization of the method through the internet and the media.[25] In a period of less than three months in 2008, there were 220 cases of attempted suicide by $H_2S$ with 208 deaths.[26] The gas is so potent it is almost immediately fatal to the person involved and potentially harmful to rescuers and bystanders.[27] In a period of 3 years, in the United States, 30 cases of suicide by $H_2S$ were identified — two cases in 2008, 10 in 2009 and 18 in 2010.[85] While the source of $H_2S$ here is not a natural one, the stories highlight the fatal nature of the gas. Indeed, more cases have now been reported in the U.S.[28]

## Health Effects and Mechanism of Action

*"Death may come on like a stroke of lightning...but usually there are first symptoms of irritation of the nervous system..."*

This was the way Alice Hamilton (1869–1970), the famed toxicologist and public health advocate from Harvard University, described the lethal effect of $H_2S$ poisoning.[29] The penetrating observation remains true today.

Hydrogen sulfide is an extremely hazardous, toxic compound and the most toxic route is through inhalation. Concentration of $H_2S$ in unpolluted ambient air is between 0.0001 and 0.0003 ppm.[30] Some asthmatics may start to have breathing problems when $H_2S$ concentration reaches 2 ppm. Workers and individuals exposed to 20 ppm or less often experience eye irritation, conjunctivitis and throat irritation. Chronic conjunctivitis may lead to complications such as visual changes and chromatic distortion and other corneal changes. At 20–100 ppm the gas causes eye and lung irritations and a number of neurological and physiological effects such as dizziness, poor memory, and irritability (Table 1). At 100 ppm and higher, the irritating effect of $H_2S$ on the mucous membrane is severe, resulting in eye and lung damages. At 500 ppm and higher, an individual may abruptly lose consciousness ("knockdown") as a consequence of respiratory paralysis, neurological effects and cellular anoxia. Death ensues if the period of exposure is extended to hours. A person exposed to 1000 ppm $H_2S$ may stop breathing after one or two breath followed by immediate collapse and death (Table 1). If the exposure is transient, as it often is in the case of oil field workers encountering a sudden air burst during drilling, it may be quickly reversible. However, neurological and chronic brain injury can sometimes follow a "knockdown".[31,32]

A potent toxic property of $H_2S$ is that it has an extremely steep exposure-response curve.[33,34,82] In other words, just a minor increase in concentration can cause a huge increase in response and death can come within minutes when high enough concentration is reached.[35] Another highly dangerous property of $H_2S$ at high concentration (>100 ppm) is that it causes olfactory paralysis: olfactory nerves are damaged to such an extent that the sense of smell is lost completely and the odor of $H_2S$ is no longer detected. Numerous deaths due to accidental or occupational exposure are associated with these two dangerous properties. In summary, as are identified by Guidotti, four toxic effects characterize $H_2S$ poisoning are: "knockdown", pulmonary edema, conjunctivitis and olfactory paralysis.

There are few studies on the health effects following chronic exposure to low levels of $H_2S$ in the environment. A study of inhabitants of Roturua, New Zealand who are exposed to as high as 1 ppm $H_2S$ from geothermal sources indicate a trend for exposure-response related to nervous, respiratory and cardiovascular diseases.[36] A study of adults living in three communities with elevated $H_2S$ concentrations originating from multiple sources reported significant increases in impaired neurobehavioral functions after exposure to $H_2S$.[37]

The multiple toxic effects of $H_2S$ suggest that the underlying mechanisms of actions are complex. Because $H_2S$ binds to cytochrome c, a key enzyme in the respiratory chain, it has been generally assumed that cellular anoxia is one of the basic mechanisms of toxicity.[38,82] The stimulatory effect of $H_2S$ on the carotid body, chemosensors associated with ventilation control, has been proposed as the main causes of initial hyperpnea (rapid, deep breathing).[39] A role of $H_2S$ on synaptic and membrane properties of brain neurons has been proposed in the rapid cessation of respiratory drive following acute exposure to high concentration of $H_2S$.[40] Recent studies indicated that endogenous $H_2S$ acts as a signaling molecule in the gastrointestinal, cardiovascular, inflammatory, and nervous systems.[41–43] It is clear that more research is needed to elucidate the effect of $H_2S$ exposure on physiological/biochemical functions before we can fully understand the toxic mechanisms.

## Adverse Effect of $H_2S$ — Case Reports

Despite an increased awareness of the toxicity of $H_2S$, reports of injury or death due to exposure to natural sources of $H_2S$ such as sewer gas, manure gas, geothermal gas persist even today. What is surprising is that many reports of fatality still originated from industrialized countries. In the U.S. for example, sewer gas, mine gas and methane collectively represent the third leading cause of fatal occupational inhalation.[44] In 2007, the U.S. Occupational Safety and Health Administration documented 13 asphyxiation deaths attributed to $H_2S$.[45] The gravity of the situation is illustrated by the following case reports.

*Sewer Gas.* In June 2009, three people from a sewer company perished while unclogging a drain at a waste transfer station near New York City.[46] A worker lost his footing and fell into an 18-foot deep dry hole. His father who was the owner of the company climbed down the hole in an attempt to rescue him, but was overcome by toxic fumes. When he failed to emerge another worker climbed down the hole and also did not survive. Fire officials found the $H_2S$ concentration in the hole to be 200 ppm, a concentration high enough to cause respiratory distress, lung damage and death if exposure is prolonged (Table 1).

Another case involving an accidental mixing of industrial chemicals in a sewer system occurred in a paper mill in Alberta, Canada.[47] Sodium hydrosulfide, a chemical used for treating wood chips, was accidently flushed into the sewer system where it reacted with sulfuric acid, which was also added to treat the mill effluent. The reaction produced $H_2S$ which leaked through a manhole and reached a group of 10 workers nearby. Three workers were overcome almost immediately. Instead of evacuating the area, three of the remaining workers attempted to drag the fallen men to fresh air without putting on respiratory equipments. Two of them passed out in the course of assisting the others. In all, two workers died and eight were injured. In addition six paramedics who transported the victims later reported symptoms consistent with $H_2S$ exposure. These tragic incidents not only illustrates the deadly nature of sewer gas, but also the commonly encountered "co-worker" effect[48,81] where colleagues trying to help also fall victim to the toxic fumes. Deaths involving worker exposed to sewer gas were also reported in other countries including France, Korea, and Spain.[49–51]

***Manure Gas.*** Death due to inhalation of manure gas occurs most commonly in farming communities. A German report[29] reviewed four such cases that resulted in 10 fatalities. In one of the cases, a farmer entered a manure pit and collapsed before he could grab hold of the safety rope. A worker outside tried to rescue him and also collapsed after entering the pit. The mother of the farmer found the victims, and attempted to help but also succumbed to the toxic fumes. Two hours later the son of the farm worker arrived at the pit and lost consciousness at the entrance to the pit. He was found alive but died later at the hospital. During autopsy, the forensic pathologists and police officers all suffered from a mild form of $H_2S$ intoxication because the autopsies were carried out in an insufficiently ventilated room. Thus, the toxic effect of manure gas can be felt beyond its immediate location. Many countries with extensive farming activities have reported cases of manure gas-related deaths,[52,53] some as recent as 2011.[54,55]

*Sour Gas.* Hydrogen sulfide in "sour" gas is a major hazard in oil and gas production. On 23 December 2003, a massive sour gas blow-out occurred at Kai County, Chongqing, China. Up to 243 people died and 10,000 people were evacuated.[56] This accident illustrates the potential deadly consequence of a sour gas accident. For many decades, poisoning by $H_2S$ has been a concern of public health agencies and the gas and petroleum industry. W.W. Burnett and colleagues[57] reported that, between 1969 and 1973, 220 cases of occupational exposure to $H_2S$ occurred in Alberta, Canada, with a mortality rate of 6%. Gas plants, pumping stations and oil rigs constitute the vast majority of the sites where such exposures occur. A review of the U.S. Occupational Safety and Health Administration investigation record for the period 1984–1994 appeared to present a higher mortality rate[58] — 80 fatalities from $H_2S$ in 57 incidents. In these cases there were 19 additional fatalities and 36 injuries among co-workers attempting to rescue the fallen. The authors suggested that fatalities demonstrated a distribution consistent with sour gas production in the Rocky mountain region. In 2006, Timothy Lambert and co-workers[59] reviewed the literature on the effects of $H_2S$ and sour gas on the eyes concluded effects are unequivocally elicited well below 100 ppm and that 0.025 ppm appears to be the lowest concentration that cause eye irritation during short-term exposure.

*Geothermal Gas.* The vast majority of the population worldwide is not affected by geothermal emissions; but for people living near areas with thermal fissures and sulfur hot springs, $H_2S$ exposure is a real risk. In Rotorua, New Zealand, a city situated in an actively degassing geothermal field, there was concern about the high concentration of $H_2S$ both indoors and outdoors.[60] A preliminary study indicated that residents were at higher risk of developing non-infectious respiratory diseases. When analysis was done on the clusters with high relative risks, they were found to be spatially coincident with the geothermal fields.[61] Two separate deaths in Rotorua hot springs have been reported. Both victims were elderly and the cause of death was determined to be $H_2S$ exposure.[62]

In Turkey, a young couple were found unconscious in their hotel room which has a thermal spring bath. The man survived while the women died at the hospital.[63] The diagnosis was acute $H_2S$ intoxication. In Japan, the death of a man working in a geothermal power plant was attributed to $H_2S$ gas poisoning.[64] His blood thiosulfate concentration, a biomarker of sulfide exposure, was at least 48 times the control level. The problem of "co-worker effects" is also evident in accidents involving geothermal gases.[65] A worker descended a hot spring reservoir for maintenance work and immediately collapsed. Four workers who were outside descended one-by-one down the manhole in a rescue effort and all collapsed. Another worker was also on his way down but noticed the strong odor. He retreated and managed to call for help after a 15 minute recovery period. The first man died and the four co-workers recovered after treatment.

## Practical Points

Hydrogen sulfide exposure is predominantly an industrial and occupational concern. There are many good guidance documents for companies and workers to consult. Booklets and bulletins on the safe handling of sewer gas, manure gas and sour gas are readily available on many government websites.[66-68] The safety procedures are summarized below:

- Obtain permit and conduct assessment of the work site.
- Consider carrying out the work without entering confined space.
- Monitor the atmosphere for toxic gases at the work site constantly.
- Wear proper breathing apparatus.
- Remove waste, silts and sludge from work site with mechanical or vacuum devices if possible.
- Ventilate the work site continuously.
- Rescue procedure must be in place.
- Work on a buddy system with workers stationed outside the work area.
- Avoid ignition sources; $H_2S$ mixture can be explosive.

- Create checklist of emergency equipment such as flashlight, wrench, breathing apparatus and resuscitator.
- Do not rely on sense of odor to assess presence of $H_2S$.
- Do not enter an accident site for rescuing purpose without proper breathing apparatus and support.

## Methane

Methane with the chemical formula of $CH_4$ is the simplest alkane. It is an odorless, colorless, and extremely flammable gas. Methane is lighter than air and will rise to the top of a confined space. Methane is the major component of natural gas; the "natural gas" odor is actually an artificial odorant which is added for safety purpose. Methane can be formed by the decay of organic materials and is common in landfills, marshes, septic systems and sewers. Methane naturally occurs in underground gas reservoirs, oil wells, and coal mines. Gas hydrate found in the deep ocean floor is an ice-like structure rich in $CH_4$. Methane in gas hydrate is a potential energy source that can be harvested, but there is concern about its effect as a greenhouse gas.[69] In humans, $CH_4$ is produced through anaerobic fermentation of carbohydrates by enteric microflora. About 30–62% of healthy human subjects are methane producers.[70]

Methane is not utilized in humans and is considered biologically inert as far as health and physiological functions are concerned. However, a recent review suggests that endogenous $CH_4$ may play a role in intestinal motor function.[70] High levels of $CH_4$ in the atmosphere, however, pose an immediate danger to humans—asphyxiation and explosion.

*Methane as an asphyxiant.* Methane is trapped in coal seams and surrounding rock strata and is released during mining activities. It can accumulate in the mine if ventilation is not adequate. Methane concentration can also rise quickly in a "gas outburst" or "gas spurt" when large volumes of $CH_4$ accumulated in the coal bed under high pressure are suddenly released into the pit. When large volumes of $CH_4$ are present in a confined space, oxygen is displaced and, depending on the extent of oxygen deficiency, effects ranging from increased pulse to loss of consciousness may occur (see Table 2, Chapter 8, Carbon Dioxide and Oxygen). Autopsy study of victims from Japanese coal mine explosions that occurred from 1965–1978 indicated that some deaths were due to asphyxiation as a result of a gas spurt.[71] On October 16, 1981, a $CH_4$ outburst occurred in a coal mine in Hokkaido, Japan. Ninety-three miners perished. Autopsies on 22 of the victims showed that the cause of deaths was asphyxiation by $CH_4$.[72] In the U.S., during June and July 1989, a total of seven farm workers in two separate incidents died after they were asphyxiated by methane gas in manure pits.[73] However, in these cases, it is not clear to what extent $H_2S$, which is also present in manure gas, was involved in the deaths.[80]

***Methane in coal mine explosion.*** Methane is produced during the process of coal formation. Some of the gas in deeper layers is adsorbed onto the coal or trapped under pressure within cracks and cavities. Mining activities can disturb these containments allowing $CH_4$ to be released into the air current. When the concentration of $CH_4$ in the air reaches 5% or higher the mixture becomes flammable — the much feared "fire damp" in coal mining jargon.[20] The cause of coal mine explosion is complex. Methane is of course a main culprit. However, coal dust level has been considered by some as equally, if not the most important, cause of mine explosion.[74] In many cases, a potent combination of both flammable methane and coal dust appeared to be the cause of explosion.[75] The inquiry into the 1992 Westray mine explosion in Nova Scotia, Canada, clearly indicated that the initial explosion was due to ignition of methane, followed by a full-blown coal dust explosion.[76] From 1900 to 2006, the number of workers who died in underground coal mine in the U.S. was 11,606 with close to 90% of the deaths being related to explosions.[77] Today, China has the highest coal mine casualties in the world. Fatality peaked in 2002 at 6,995 death and gradually decreased to 2,433 deaths by 2010.[78,79] Gas explosion accounts for close to 45% of the death from 2001–2010.[78] Mining deaths due to methane explosion remains a concern in China and many other countries and further improvement in safety procedure is needed.

## Practical Points

- Manufacturing and resource industries must adhere to the appropriate industry and government guidelines on methane.
- Methane may be present in manure pits and cause asphyxiation. The work area must be well-ventilated and co-workers must be present outside to help in emergency.
- Individuals using methane should ensure that an ignition source is not present and that the space is well-ventilated.

## References

1. U.S. EPA. (2008) Integrated science assessment for sulfur oxides — health criteria. EPA/600/R-08/047F. National Center for Environmental Assessment, Office of Research and Development, U.S. Environmental Protection Agency, Research Triangle Park, NC. [http://cfpub.epa.gov.proxy.bib.uottawa.ca/ncea/isa/recordisplay.cfm?deid=198843].

2. Toxicological Profile for Sulphur Dioxide. (1998) Agency for Toxic Substance and Disease Registry. U.S. Department of Health and Human Services. Atlanta, Georgia.

3. World Health Organization. (2006) *WHO Air Quality Guidelines for Particulate Matter, Ozone, Nitrogen Dioxide and Sulfur Dioxide.* Global update 2005, Summary of risk assessment, World Health Organization, Geneva, Switzerland. [http://whqlibdoc.who.int/hq/2006/WHO_SDE_PHE_OEH_06.02_eng.pdf].

4. Symonds RB, Rose WI, Bluth G, Gerlach TM *et al.* Volcanic gas study: Methods, results and applications. In: Carroll MR, Holloway JR (eds.), *Volatiles in Magma,* Vol. 30, pp. 1–66. Mineralogical Society of America Reviews in Mineralogy.

5. National Park Service. *Hawai'i Volcanoes National Park. Sulfur Dioxide (SO₂) Advisory Program.* U.S. National Park Service. [http://www.nature.nps.gov/air/webcams/parks/havoso2alert/havoadvisories.cfm].

6. McGillivray K. The Smoking Hill. Canadian Broadcasting Corporation. [http://www.cbc.ca/sevenwonders/more_wonders_northwest_territories.html].

7. Freedman B, Sobers V, Hutchinson, TC, Gazing WI *et al.* (1990) Intense, natural pollution affects arctic tundra vegetation at the smoking hills, Canada. *Ecology* 7: 492–503.

8. Small D, Neumann T. (2001) The global distribution of human population and recent volcanism. *Environ. Haz.* 3: 93–109.

9. Thordarson T, Self S. (2003) Atmospheric and environmental effects of the 1783–1784 Laki eruption: A review and reassessment. *J. Geophys. Res.* 108: D1.

10. Schmidt A, Ostro B, Carslaw KS *et al.* (2011) Excess mortality in Europe following a future Laki-style Icelandic eruption. *Proc. Natl. Acad. Sci. U.S.A.* 108: 15710–15715.

11. Grattan J, Durand M, Taylor S. (2003) Illness and elevated human mortality in Europe coincident with the Laki Fissure eruption. *Geological Society, London, Special Publications* 213: 401–414.

12. Witham CS, Oppenheimer C. (2004) Mortality in England during the 1783–1784 Laki Craters eruption. *Bull. Volcanol.* 67: 15–26.

13. Ishigami A, Kikuchi Y, Iwasawa S *et al.* (2008) Volcanic sulphur dioxide and acute respiratory symptoms on Miyakejima Island. *Occup. Environ. Med.* 65: 701–707.

14. Iwaswa S, Kikuchi Y, Nishiwaki Y *et al.* (2009) Effects of $SO_2$ on respiratory system of adult Miyakejima residents 2 years after returning to the island. *J. Occup. Health.* 51: 38–47.

15. Longo BM, Yang W, Green JB *et al.* (2010) Acute health effects associated with exposure to volcanic air pollution (vog) from increased activity at Kilauea Volcano in 2008. *J. Toxicol. Environ. Health A* 73: 1370–1381.

16. Longo BM, Yang W, Green JB *et al.* (2010) An indoor air quality assessment for vulnerable populations exposed to volcanic vog from Kilauea Volcano. *Fam Community Health* 33: 21–31.

17. U.S. Department of Health and Human Services. (2006) *Toxicological Profile for Hydrogen Sulfide*, Agency for Toxic Substance and Disease Registry, U.S. Department of Health and Human Services, Atlanta, Georgia.

18. WHO. (2003) *Concise International Chemical Assessment Documents 53. Hydrogen Sulfide: Human Health Aspects.* World Health Organization, Geneva. Report.

19. Lambert TW, Goodwin VM, Stefani D, Strosher L *et al.* (2006) Hydrogen sulfide (H$_2$S) and sour gas effects on the eye. A historical perspective. *Sci. Total Environ.* **367**: 1–22.

20. Unwin ID. Mine monitoring for safety and health. [http://www.angel-fire.com/mech/ians_ coal_page/Health_and_safety_2007.pdf].

21. British Broadcasting Corporation. [http://news.bbc.co.uk/onthisday/hi/dates/stories/December/30/newsid_ 2547000/2547587.stm].

22. Wells RE, Slocombe RF, Trapp AL. (1982) Acute toxicosis of budgerigars (Melopsittacus undulatus) caused by pyrolysis products from heated polytetrafluoroethylene: Clinical study. *Am. J. Vet. Res.* **43**: 1238–1242.

23. Blandford TB, Seamon PJ, Hughes R *et al.* (1975) A case of polytetra-fluoroethylenepoisoning in cockatiels accompanied by polymer fume fever in the owner. *Vet. Rec.* **96**: 175–178.

24. Stotz JH, Galey F, Johnson B. (1992) Sudden death in ten psittacine birds associated with the operation of a self-cleaning oven. *Vet. Hum. Toxicol.* **34**: 420–421.

25. Maebashi K, Iwadate K, Sakai K *et al.* (2011) Toxicological analysis of 17 autopsy cases of hydrogen sulfide poisoning resulting from the inhalation of intentionally generated hydrogen sulfide gas. *Forensic Sci. Int.* **207**: 91–95.

26. Morii D, Miyagatani Y, Nakamae N *et al.* (2010) Japanese experience of hydrogen sulfide: The suicide craze in 2008. *J. Occup. Med. Toxicol.* **29**: 5–28.

27. Truscott A. (2008) Suicide fad threatens neighbours, rescuers. *Can. Med. Assoc. J.* **179**: 312–313.

28. Reedy SJ, Schwartz MD, Morgan BW. (2011) Suicide fads: Frequency and characteristics of hydrogen sulfide suicides in the United States. *West J. Emerg. Med.* **12**: 300–304.

29. Oesterhelweg L, Püschel K. (2008) Death may come on like a stroke of lightening: Phenomenological and morphological aspects of fatalities caused by manure gas. *Int. J. Legal Med.* **122**: 101–107.

30. US EPA. Sulfur Dioxide (SO$_2$) Primary Standards–Table of Historical SO$_2$ NAAQS. [http://www.epa.gov/ttn/naaqs/standards/so2/s_so2_history.html].

31. Tvedt B, Skyberg K, Aaserud O *et al.* (1991) Brain damage caused by hydrogen sulfide: A follow-up study of six patients. *Am. J. Ind. Med.* **20**: 91–101.

32. Guidotti TL. (2010) Hydrogen sulfide: Advances in understanding human toxicity. *Int. J. Toxicol.* **29**: 569–581.

33. Prior MG, Sharma AK, Yong S, Lopez A *et al.* (1988) Concentration-time interactions in hydrogen sulphide toxicity in rats. *Can. J. Vet. Res.* **52**: 375–379.

34. Brown KG, Strickland JA. (2003) Utilizing data from multiple studies (meta-analysis) to determine effective dose-duration levels. Example: Rats and mice exposed to hydrogen sulfide. *Regul. Toxicol. Pharmacol.* **37**: 305–317.

35. Guidotti TL. (1996) Hydrogen sulphide. *Occup. Med.* **46**: 367–371.

36. Bates MN, Garrett N, Shoemack P. (2002) Investigation of health effects of hydrogen sulfide from a geothermal source. *Arch. Environ. Health.* **57**: 405–411.

37. Kilburn KH, Thrasher JD, Gray MR. (2010) Low-level hydrogen sulfide and central nervous system dysfunction. *Toxicol. Ind. Health.* **26**: 387–405.

38. Beauchamp RO Jr, Bus JS, Popp JA *et al.* (1983) A critical review of the literature on carbon disulfide toxicity. *Crit. Rev. Toxicol.* **11**: 169–278.

39. Ammann HM. (1986) A new look at physiological respiratory response to hydrogen sulfide poisoning. *J. Haz. Mater.* **13**: 369–374.

40. Toxicological Profile for Hydrogen Sulfide. (2006) Agency for Toxic Substance and Disease Registry. U.S. Department of Health and Human Services. Atlanta, Georgia.

41. Wagner F, Asfar P, Calzia E *et al.* (2009) Bench-to-bedside review: Hydrogen sulfide–the third gaseous transmitter: Applications for critical care. *Crit. Care* **13**: 213.

42. Tang G, Wu L, Wang R. (2010) Interaction of hydrogen sulfide with ion channels. *Clin Exp. Pharmacol. Physiol.* **37**: 753–763.

43. Wallace JL. (2010) Physiological and pathophysiological roles of hydrogen sulfide in the gastrointestinal tract. *Antioxid. Redox Signal.* **12**: 1125–1133.

44. Valent F, McGwin G Jr, Bovenzi M, Barbone F *et al.* (2002) Fatal work-related inhalation of harmful substances in the United States. *Chest.* **121**: 969–975.

45. Ballerino-Regan D, Longmire AW. (2010) Hydrogen sulfide exposure as a cause of sudden occupational death. *Arch. Pathol. Lab Med.* **134**: 1105.

46. Associated Press Release, June 29, 2009. Frank Eltman.

47. Khoshniat H. (2008) H₂S: The silent killer. *Occup. Health Saf.* **77**: 55.

48. Fuller DC, Suruda AJ. (2000) Occupationally related hydrogen sulfide deaths in the United States from 1984 to 1994. *J. Occup. Environ. Med.* **42**: 939–942.

49. Christia-Lotter A, Bartoli C, Piercecchi-Marti MD *et al.* (2007) Fatal occupational inhalation of hydrogen sulfide. *Forensic Sci. Int.* **169**: 206–209.

50. Nogué S, Pou R, Fernández J, Sanz-Gallén P *et al.* (2011) Fatal hydrogen sulphide poisoning in unconfined spaces. *Occup. Med.* **61**: 212–214.

51. Lee EC, Kwan J, Leem JH *et al.* (2009) Hydrogen sulfide intoxication with dilated cardiomyopathy. *J. Occup. Health* **51**: 522–525.

52. Knoblauch A, Steiner B. (1999) Major accidents related to manure: A case series from Switzerland. *Int. J. Occup. Environ. Health* **5**: 177–186.

53. Belley R, Bernard N, Côté M *et al.* (2005) Hyperbaric oxygen therapy in the management of two cases of hydrogen sulfide toxicity from liquid manure. *Can. J. Emerg. Med.* **7**: 257–61.

54. Zaba C, Marcinkowski JT, Wojtyła A *et al.* (2011) Acute collective gas poisoning at work in a manure storage tank. *Ann. Agric. Environ. Med.* **18**: 448–451.

55. Knubben-Schweizer G, Brosinski K, Steiner B. (2011) Hydrogen sulfide poisoning (manure gas poisoning) in a cattle facility with open stall design. *Schweiz. Arch. Tierheilkd.* **153**: 127–129.

56. MA Q, Zhang L. (2011) CFD simulation study on gas dispersion for risk assessment: A case study of sour gas well blowout. *Safety Sci.* **49**: 1289–1295.

57. Burnett WW, King EG, Grace M, Hall WF *et al.* (1977) Hydrogen sulfide poisoning: Review of 5 years experience. *Can. Med. Assoc. J.* **17**: 1277–1280.

58. Fuller DC, Suruda AJ. (2000) Occupationally related hydrogen sulfide deaths in the United States from 1984 to 1994. *J. Occup. Environ. Med.* **42**: 939–942.

59. Lambert TW, Goodwin VM, Stefani D, Strosher L *et al.* (2006) Hydrogen sulfide (H₂S) and sour gas effects on the eye. A historical perspective. *Sci. Total Environ.* **367**: 1–22.

60. Durand M, Scott BJ. (2005) Geothermal ground gas emissions and indoor air pollution in Rotorua, New Zealand. *Sci. Total Environ.* **345**: 69–80.

61. Durand M, Wilson JG. (2006) Spatial analysis of respiratory disease on an urbanized geothermal field. *Environ. Res.* **101**: 238–245.

62. Bassindale T, Hosking M. (2011) Deaths in Rotorua's geothermal hot pools: Hydrogen sulphide poisoning. *Forensic Sci. Int.* **207**: e28–e29.

63. Daldal H, Beder B, Serin S, Sungurtekin H *et al.* (2010) Hydrogen sulfide toxicity in a thermal spring: A fatal outcome. *Clin. Toxicol (Phila)* **48**: 755–756.

64. Kage S, Ito S, Kishida T *et al.* (1988) A fatal case of hydrogen sulfide poisoning in a geothermal power plant. *J. Forensic Sci.* **43**: 908–910.

65. Deng JF, Chang SC. (1987) Hydrogen sulfide poisonings in hot-spring reservoir cleaning: Two case reports. *Am. J. Ind. Med.* **11**: 447–451.

66. Alberta. Workplace Health and Safety Bulletin. Hydrogen Sulphide at the Work Site. Government of Alberta. Canada. [http://employment. alberta.ca/documents/WHS/WHS-PUB-CH029.pdf].

67. Saskatchewan. Hydrogen Sulfide-The Deadliest Manure Gas, Occupational Health and Safety Division. Department of Labour. Saskatchewan, Canada. [http://www.lrws.gov.sk.ca/hydrogen-sulfide-deadliest-manure-gas].

68. Hong Kong. (2007) Prevention of Gas Poisoning in Drainage Work, The Occupational Safety and Health Branch, Labour Department. Hon Kong, SAR. [http://www.labour. gov.hk/eng/public/oh/Drainage.pdf].

69. Kvenvolden KA. (1999) Potential effects of gas hydrate on human welfare. *Proc. Natl. Acad. Sci. U.S.A.* **96**: 3420–3426.

70. Sahakian AB, Jee SR, Pimentel M. (2010) Methane and the gastrointestinal tract. *Dig. Dis. Sci.* **55**: 2135–2143.

71. Suzutani T, Ishibashi H, Takatori T. (1979) Medico-legal studies on the deaths from coal-mine accidents. 3. Causes of death (author's transl). *Hokkaido Igaku Zasshi* **54**: 479–486.

72. Terazawa K, Takatori T, Tomii S, Nakano K *et al.* (1985) Methane asphyxia. Coal mine accident investigation of distribution of gas. *Am. J. Forensic Med. Pathol.* **6**: 211–214.

73. Fatalities attributed to methane asphyxia in manure waste pits–Ohio, Michigan. (1989) *MMWR Morb Mortal Wkly Rep* **38**: 583–586.

74. Aldrich M. (1995) Preventing the needless peril of the coal mine: The bureau of mines and the campaign against coal mine explosions, 1910–1940. *Technol. Cult.* **3**: 483–518.

75. Amyotte PR, Eckhoff RK. (2010) Dust explosion causation, prevention and mitigation: An overview. *J. Chem. Health Saf.* **17**: 15–28.

76. Richard, KP. (1997) *The Westray Story — A Predictable Path to Disaster, Report of the Westray Mine Public Inquiry.* Province of Nova Scotia: Halifax, NS, Canada. Report.

77. Underground coal mining disasters and fatalities–United States, 1900–2006. (2009) *Morb Mortal Wkly Rep* **57**: 1379–1383.

78. Chen H, Qi H, Long R, Zhang M *et al.* (2012) Research on 10-year tendency of China coal mine accidents and the characteristics of human factors. *Safety Sci.* **50**: 745–750.

79. He X, Song L. (2012) Status and future tasks of coal mining safety in China. *Safety Sci.* **50**: 894–898.

80. Gargi J, Thind AS. (1993) Death scene gas analysis in suspected methane asphyxia. *Am. J. Forensic Med. Pathol.* **14**: 350–351.

81. Hendrickson RG, Chang A, Hamilton RJ. (2004) Co-worker fatalities from hydrogen sulfide. *Am. J. Ind. Med.* **45**: 346–350.

82. Woodall GM, Smith RL, Granville GC. (2005) Proceedings of the hydrogen sulfide health research and risk assessment symposium October 31–November 2, 2000. *Inhal. Toxicol.* **17**: 593–639.

83. Longo BM, Rossignol A, Green JB. (2008) Cardiorespiratory health effects associated with sulphurous volcanic air pollution. *Public Health* **122**: 809–820.

84. Longo BM, Yang W. (2008) Acute bronchitis and volcanic air pollution: A community-based cohort study at Kilauea Volcano, Hawai'i, U.S.A. *J. Toxicol. Environ. Health A* **71**: 1565–1571.

85. Reed SJD, Schwartz MD, Morgan BW. (2011) Suicide fades: Frequency and characteristics of hydrogen sulphide suicide in the United States. *Western J. Emerg. Med.* **12**: 300–304.

86. Johns DO, Svendsgaard D, Linn WS. (2010) Analysis of the concentration-respiratory response among asthmatics following controlled short-term exposures to sulfur dioxide. *Inhal. Toxicol.* **22**: 1184–1193.

87. Johns DO, Linn WS. (2011) A review of controlled human $SO_2$ exposure studies contributing to the US EPA integrated science assessment for sulfur oxides. *Inhal. Toxicol.* **23**: 33–43.

88. Mathews WH, Bustin RM. (1984) Why do the smoking hills smoke? *Can. J. Earth. Sci.* **21**: 737–742.

# CHAPTER 10

# HARDNESS OF WATER

For people who rely on the municipal or in-house water treatment system, water hardness may not be an issue. However, for people living in a rural area or in a less-developed country, water hardness may be a concern. It is estimated that over 1.1 billion people around the world lack access to improved water supply.[1]

The principal sources of hardness in water are dissolved polyvalent metal ions originated from sedimentary rocks, and seepage and runoffs from soil. We recognize water as hard when more soap is needed to form lather and when the soap forms a scum in and around the container. This happens because hard water contains higher concentrations of polyvalent metallic ions, mainly calcium and magnesium, which react with fatty acids in soap to form insoluble precipitates. Another well-known problem with hard water is the deposition of minerals that causes scaling in containers, water heaters, and pipes. In chemical terms, water hardness is defined as the molar sum of calcium and magnesium found in water and is expressed as the concentration of calcium carbonate ($CaCO_3$) equivalent. Water containing calcium carbonate at concentration below 60 mg/L is generally considered as soft; 60–120 mg/L as moderately hard, 120–180 mg/L as hard, and >180 mg/L, very hard. Calcium concentrations up to and exceeding 100 mg/L are common in natural water, particularly groundwater. Magnesium is present in natural groundwater usually at 0–50 mg/L but rarely above 100 mg/L.[2,3]

Unlike aluminum, lead and mercury, which serve no known physiological functions (see Chapters 1, 5 and 6), calcium and magnesium are both essential minerals for our body. Calcium is the predominant constituent of bone and teeth and other calcium-containing tissues. Normal calcium metabolism is essential for healthy bone growth. Muscle contraction is mediated by the movement of intracellular calcium. Calcium is critical in many biochemical/physiological processes, such as secretion, neurotransmission, electrical conduction of the heart, mitochondrial functions, and control of cell death. Deficiency in calcium intake has been associated with increased risk of osteoporosis, nephrolithiasis, colorectal cancer, hypertension and stroke, coronary artery disease, insulin resistance and obesity.[3-5] Magnesium participates in many cellular functions, e.g. transport of potassium and calcium ions, modulation of signal transduction, energy metabolism and cell proliferation. Magnesium is a cofactor in many enzyme catalyzed reactions. For example, important enzyme reactions involving adeonosine triphosphate (ATP) invariably require magnesium as a cofactor. Magnesium involvement is essential in protein and nucleic acid synthesis, cell cycle processes, cytoskeletal and mitochondrial integrity and for the binding of biological substances to the plasma membrane. Magnesium modulates cellular ion transport by pumps, carriers and channels and thereby influences signal transduction and the cytosolic concentrations of calcium and potassium. Low magnesium level has been associated with endothelial dysfunction and increased vascular reaction. Magnesium deficiency has been implicated in the pathogenesis of hypertension, cardiac arrhythmia and the development of coronary artery disease.[3,6,7] There is experimental evidence that low serum magnesium levels are associated with increased risk of inflammation, thrombus formation, and atherogenesis, whereas normal magnesium levels protect the endothelium by enhancing synthesis of nitric oxide, a potent vasodilator.[8,9] Magnesium sulfate administration is a standard therapy for eclampsia and preeclampsia.[44] Both oral or intravenous administrations of magnesium have been employed for the treatment of certain types of arrhythmia.[45]

Although calcium and magnesium are essential minerals for our bodies, their presence in elevated levels in hard waters elicits many questions from the general population about their potential health effects. The most frequently asked questions concerns kidney stones, heart diseases and eczema. The following sections will attempt to address these questions.

## Does Hard Water Increases Risk of Urinary Stones?

Urinary stones have been a problem for humans since ancient times; a 5,000-year-old bladder stone was uncovered at the Egyptian funeral site of El Amrah.[10] The ancient Indian text Sushruta Samhita (600 BCE to 600 AD) contained a detail description of a perirectal operation to remove urinary stones collected in the bladder. The same procedure appeared to have been adopted and modified by the Roman physician Aulus Cornelius Celsus (*ca.* 25 BC–*ca.* 50 AD) who also described a number of instruments used in the surgical procedure.[11]

Urinary stone disease (urolithiasis), is also one of the most common afflictions of modern society.[12] About 80% of patients with the disease have stones that are composed of calcium in combination with oxalate; the remaining is composed of calcium phosphate, uric acid and struvite (infection related stones). It is therefore not surprising that the public has the perception that calcium-rich hard water contributes to urinary stone formation. However, epidemiological studies appear to show that the opposite may be true.

Early studies examining incidence of urinary stones in areas with soft and hard water suggested that hard water was, in fact, associated with a low incidence of stones.[13,14] Churchill and co-workers[15] conducted a study in Newfoundland, Canada, which had the highest hospitalization rate for urinary stones in North America, and found no association between district prevalence and the mean district drinking-water hardness. R. Sirakowski *et al.*[16] examined discharge data from United States hospitals and found a negative correlation between urinary stones and reported water hardness. P.P Singh and R. Kiran[17] examined drinking-water from 59 localities of Udaipur region of India and found that the highest prevalence of urolithiasis was in localities where the water was the softest. Michelle Lopez and Bernd Hoppe[18] examined the global diversity in incidences of urinary stones. They suggested that racial distributions, social economic status, and changing dietary habits were the main contributing factors. Water hardness was noticeably not included as a factor.

It should be noted that these geographically-based studies have their limitations. Because aggregate data were used, the studies cannot detect individual effects. Study results are also compromised by important confounding factors such as dietary habits, physical activities and underlying diseases. Nevertheless, most of the epidemiological studies carried out in the past 50 years do not support the popular perception that hard water increases the risk of developing urinary stones. There are, however, a few studies that do not agree with this conclusion. For example, G. Coen *et al.*[19] reported that an increase in drinking water hardness results in an increase in urinary stone incidence. A World Health Organization review[3] concluded that dietary calcium decreases the risk of kidney stones by binding oxalic acid. However, oral calcium and calcium supplements not taken together with food may increase the risk of kidney stones.

Although public concern is focused on water hardness, there are in fact many other risk factors for urinary stones. These factors include family history, gender, age, sedentary living style, body mass index, fluid intake, dietary factors (protein, oxalate, sucrose, ascorbic acid), metabolic disorders, genetic determinants, air temperature, and even sunlight.[18,20–23]

In contrast to calcium, the potential role of magnesium, the other major constituent of hard water, has received much less attention. Japanese studies reported that magnesium–calcium ratio in tap water correlated negatively with the incidence of urolithiasis.[24] A recent study proposed the use of a "stone risk index" which took into consideration the concentration of calcium, magnesium and bicarbonate in drinking-water. It concluded that incidence of urinary stone was inversely correlated with magnesium concentration in drinking water.[25] Magnesium oxide has been used experimentally as a prophylaxis to reduce the risk of kidney stones.[26,27]

In summary, the majority of studies did not support the popular conception that hard water increases the risk of uriolithiasis.

However, a definitive study has not yet been done. Such a study must address issues that include magnesium, calcium and bicarbonate effects, biological and genetic risk factors, life style, socio-economic variables, and dietary habits.

## Does Hard Water Cause or Exacerbate Eczema?

Eczema (atopic dermatitis) is a common skin disease characterized by recurring episodes of itching and inflammation and a chronic, relapsing course. In industrial countries, about 15–30% of children and 2–10% of adults are affected by this condition.[28,29] Hard water has long been suggested to be somehow involved in the development of eczema but only anecdotal cases have been presented. In 1998, N.J. McNally and co-worker[30] published the result of an ecological study indicating that exposure to hard water in British homes increased the risk of eczema in children of primary school age but not children of secondary school age. They suggested that hard water may be linked to eczema in two ways. (1) Calcium and magnesium may act as a direct irritant or may modify the effects of other chemicals, and (2) A need for more soap and shampoo to obtain a lather may produce an irritant on the skin and cause eczema in predisposed children. Eczema is a concern in Japan as the prevalence of eczema of Japanese children aged 6–7 years is the second highest in 56 countries. A study was conducted in which 458,284 Japanese children aged 6–12 years of age were exposed to five categories of water hardness. An association between water hardness and increased risk of eczema was observed.[31] A Spanish study[32] found that water hardness was significantly associated with the development of eczema in 6–7 year olds but not in 13–14 year old schoolchildren, thus in agreement with the British study. Based on these findings, Thomas and co-workers[33] initiated a randomized control trial to investigate whether soft water (from installed ion-exchange water softener) has any positive effect in the care of children with eczema. They found that the use of water softeners did not provide additional benefits to the usual care administered to these patients. Therefore, they did not recommend the use of ion-exchange water softeners for the treatment of children with moderate to severe eczema. A recent Belgian study[46] confirmed that domestic water hardness is linearly associated with the prevalence of childhood eczema. Interestingly, swimming in chlorinated pools might also act together to increase the prevalence.

### Does Hard Water Reduce Risk of Cardiovascular Disease?

In the late1950s, Jun Kobayashi[34] was intrigued by the exceptionally high mortality rate due to cerebrovascular disease in Japan[35] and by the marked variation in the incidence of this disease from prefectures to prefectures. Further investigations revealed that there appeared to be an association between cerebrovascular death rate and the acidity of river water used for drinking. He noted that many rivers in Japan are acidic because of the low concentrations of calcium carbonate, which is mildly alkali. One of his suggestions was that high calcium carbonate in hard water was responsible for the reduction in the death rate. This study was followed by a study conducted in the United States which found significant negative correlations between water hardness and death rates from cardiovascular and cerebrovascular diseases.[36] At about the same time J.N. Morris and colleagues[37] surveyed the county boroughs of England and Wales and observed that the softer the water supply in the county boroughs, the higher was the death-rate from cardiovascular disease. They concluded that a negative correlation exists between water hardness and death rates from cardiovascular disease. In a further study, they reported that in areas where water-supply had been changed substantially, cardiovascular death rate showed a favorable effect in towns where water had become harder, and an unfavorable effect where water had become softer.[38] It should be noted that there has also been persistent reports of an absence of correlation between water hardness and death rate from cardiovascular disease.[39–41] The study by Lina Luers[41] followed a cohort of 120,852 men and women aged 55–69 years from 1986 to 1996. She found no evidence of an association between tap water hardness, magnesium or calcium concentrations, and ischemic heart disease mortality or stroke mortality. M-P Sauvant and D. Pepin[42] reviewed epidemiological studies (19 ecological, 5 cohort and 6 case-control studies) conducted from 1960–2000. They concluded that these studies as a whole pointed to a world-wide negative relationship between cardiovascular disease death rate and drinking-water hardness. However due to inherent and methodological deficiencies in these studies, they considered the observed relationship between drinking water hardness and cardiovascular death rate as indirect, and the cause was not conclusively established. Sauvant and Pepin suggested more studies, especially large intervention studies conducted through well-controlled public health programs. Regrettably, more than 10 years after their review, the scientific community is still at the stage of calling for more studies.[41,43] The World Health Organization,[3] for example, concluded that "*Although there is some evidence from epidemiological studies for a protective effect of magnesium or hardness on cardiovascular mortality, the evidence is being debated and does not prove causality*" and noted that "*further studies are being conducted*". More than half a century has passed since the Japanese study, and we are still not absolutely certain that hard water offers protection from cardiovascular diseases.

## Practical Points

There is no national or international guidance limits on water hardness that is based on health considerations. The World Health Organization[3] states that "*there are insufficient data to suggest either minimum or maximum concentrations of mineral at this time*". Given that the debate is still going on in the scientific community, the following World Heath Organization[3] statement may be the most appropriate consensus opinion as of to date:

For Urinary stone:

> *The relationship between calcium intake and the incidence of kidney stone is dependent on whether calcium is consumed with food or separately. Calcium that reaches the lower small intestine actually protects against kidney stone by binding oxalic acid. Calcium ingested from water with food*

*has the same effect. Epidemiological evidence is strong that dietary calcium reduces the incidence of kidney stone. In contrast, the result of a large randomized trial suggests an increased risk of kidney stones associated with calcium supplement, possibly because the calcium was not ingested with food or the supplement was taken by those who exceeded the upper level of 2,500 mg/day.*

For Cardiovascular disease:

*A large number of epidemiological studies have investigated the potential health effects of drinking-water hardness. Most of these have been ecological studies and have found an inverse relationship between water hardness and cardiovascular mortality. Inherent weaknesses in the ecological epidemiological design limit the conclusions that can be drawn from these studies.*

    *Based on identified case-control and cohort studies, there is no evidence of an association between water hardness or calcium and acute myocardial infarction or death from cardiovascular disease (acute myocardial infarction, stroke and hypertension). There does not appear to be an association between drinking-water magnesium and acute myocardial infarction. However the studies do show a negative association (i.e., a protective effect) between cardiovascular mortality and drinking water magnesium. Although this association does not necessarily demonstrate causality, it is consistent with the well known effects of magnesium on cardiovascular function."*

For eczema, there are no guidance documents or position statements concerning hardness of water. Based on existing evidence, it is prudent to avoid bathing young children who have eczema or are predisposed to the development of eczema in hard water.

## References

1. Global Water Supply and Sanitation Assessment 2000 Report. 2000 World Health Organization and United Nations Children's Fund. [http://www.who.int/water_sanitation_health/ monitoring/ jmp2000.pdf].
2. Schwartz BF, Bruce J, Leslie S, Stoller ML. (2001) Rethinking the role of urinary magnesium in calcium urolithiasis. *J. Endourol.* **15**: 233–235.

3. Hardness in Drinking Water. (2011) Background Document for Development of WHO Guidelines for Drinking-water Qualtiy. WHO/HSE/WSH/10.01/Rev/1World Health Organization.

4. NRC. (1989) Recommended Dietary Allowances, 10th edn. Subcommittee on the Tenth Edition of the RDAs. Food and Nutrition Board, Commission on Life Sciences, National Research Council, National Academy Press, Washington, D.C.

5. Drago I, Pizzo P, Pozzan T. (2011) After half a century mitochondrial calcium in-and efflux machineries reveal themselves. *EMBO J.* **30**: 4119–4125.

6. Saris NE, Mervaala E, Karppanen H *et al.* (2000) Magnesium. An update on physiological, clinical and analytical aspects. *Clin. Chim. Acta.* **294**: 1–26.

7. Shechter M. (2003) Does magnesium have a role in the treatment of patients with coronary artery disease? *Am. J. Cardiovasc. Drugs* **3**: 23–239.

8. Delva P. (2003) Magnesium and coronary heart disease. *Mol. Aspects Med.* **24**: 63–78.

9. Maier JAM, Malpuech-Brugere C, Zimowska *et al.* (2004) Low magnesium promotes endothelial cell dysfunction: Implications for atherosclerosis, inflammation and thrombosis. *Biochim. Biophys. Acta.* **1689**: 13–21.

10. Eknoyan G. (2004) History of urolithiasis. *Clin. Rev. Bone Min. Metab.* **2**: 177–185.

11. Shah J, Whitfield HN. (2002) Urolithiasis through the ages. *BJU Int.* **89**: 801–810.

12. Ramello A, Vitale C, Marangella M. (2000) Epidemiology of nephrolithiasis. *J. Nephrol.* **13**(Suppl 3): S45–S50.

13. Mates J. (1968) *Renal Stone Research Symposium.* In Hodgkinson and Nordin (eds.), pp. 59. Churchill Ltd, London.

14. Rose GA, Westbury EJ. (1975) The influence of calcium content of water, intake of vegetables and fruit and of other food factors upon the incidence of renal calculi. *Urol. Res.* **3**: 61–66.

15. Churchill DN, Maloney CM, Bear J *et al.* (1980) Urolithiasis — a study of drinking water hardness and genetic factors. *J. Chronic Dis.* **33**: 727–731.

16. Sierakowski R, Finlayson B, Landes R. (1979) Stone incidence as related to water hardness in different geographical regions of the United States. *Urol. Res.* **7**: 157–160.

17. Singh P, Kiran R. (1993) Are we overstressing water quality in urinary stone disease? *Int. Urol. Nephrol.* **25**: 29–36.

18. López M, Hoppe B. (2011) History, epidemiology and regional diversities of urolithiasis. *Pediatr. Nephrol.* **25**: 49–59.

19. Coen G, Sardella D, Barbera G *et al.* (2001) Urinary composition and lithogenic risk in normal subjects following oligomineral versus bicarbonate-alkaline high calcium mineral water intake. *Urol. Int.* **67**: 49–53.

20. Soucie JM, Coates RJ, McClellan W *et al.* (1996) Relation between geographic variability in kidney stones prevalence and risk factors for stones. *Am. J. Epidemiol.* **143**: 487–495.

21. Pearle MS. (2001) Prevention of nephrolithiasis. *Curr. Opin. Nephrol. Hypertens.* **10**: 203–209.

22. Curhan GC. (2007) Epidemiology of stone disease. *Urol. Clin. North Am.* **34**: 287–293.

23. Parmar MS. (2004) Kidney stones. *BMJ* **328**: 1420–1424.

24. Kohri K, Kodama M, Ishikawa Y. (1989) Magnesium-to-calcium ratio in tap water and its relationship to geological features and the incidence of calcium-containing urinary stones. *J. Urol.* **142**: 1272–1275.

25. Basiri A, Shakhssalim N, Khoshdel AR *et al.* (2011) Drinking water composition and incidence of urinary calculus: Introducing a new index. *Iran J. Kidney Dis.* **5**: 15–20.

26. Khan SR, Shevock PN, Hackett RL. (1993) Magnesium oxide administration and prevention of calcium oxalate nephrolithiasis. *J. Urol.* 412–416.

27. Kato Y, Yamaguchi S, Yachiku S *et al.* (2004) Changes in urinary parameters after oral administration of potassium-sodium citrate and magnesium oxide to prevent urolithiasis. *Urol.* **63**: 7–11.

28. Bieber T. (2008) Atopic dermatitis. *N. Engl. J. Med.* **358**: 1483–1484.

29. Ghazvini P, Pagan LC, Rutledge TK, Goodman Jr HS *et al.* (2010) Atopic dermatitis. *J. Pharm. Pract.* **23**: 110–116.

30. McNally NJ, Williams HC, Phillips DR *et al.* (1998) Atopic eczema and domestic water hardness. *Lancet* **352**: 527–31.

31. Miyake Y, Yokoyama T, Yura A *et al.* (2004) Ecological association of water hardness with prevalence of childhood atopic dermatitis in a Japanese urban area. *Environ. Res.* **94**: 33–37.

32. Arnedo-Pena A, Bellido-Blasco J, Puig-Barbera J *et al.* (2007) Domestic water hardness and prevalence of atopic eczema in Castellon (Spain) school children. *Salud Publica Mex.* **49**: 295–301.

33. Thomas KS, Koller K, Dean T *et al.* (2011) A multicentre randomised controlled trial and economic evaluation of ion-exchange water softeners for the treatment of eczema in children: The softened water eczema trial (SWET). *Health Technol. Assess.* **15**: 1–156.

34. Kobayashi J. (1957) On Geographical relationship between the chemical nature of river water and death-rate from apoplexy. *Berichte d Ohara Inst. f Landwirtsch Biologie* **11**: 12–21. [http://www.mgwater.com/story.shtml].

35. Mirzaei M, Truswell AS, Arnett K *et al.* (2012) Cerebrovascular disease in 48 countries: Secular trends in mortality 1950–2005. *J. Neurol. Neurosurg. Psychiatry* **83**: 138–145.

36. Schroeder HA. (1960) Relations between hardness of water and death rates from certain chronic and degenerative diseases in the United States. *J. Chronic Dis.* **12**: 586–591.

37. Morris JN, Crawford MD, Heady JA. (1961) Hardness of local water supplies and mortality from cardiovascular disease in the county boroughs of England and Wales. *Lancet* **1**: 860–862.

38. Morris JN, Crawford MD, Heady JA. (1962) Hardness of local water-supplies and mortality from cardiovascular disease. *Lancet* **280**: 506–507.

39. Allwright SP, Coulson A, Detels R, Porter CE *et al.* (1974) Mortality and water-hardness in three matched communities in Los Angeles. *Lancet* **1**: 860–864.

40. Morris RW, Walker M, Lennon *et al.* (2008) Hard drinking water does not protect against cardiovascular disease: New evidence from the British regional heart study. *Eur. J. Cardiovasc. Prev. Rehabil.* **15**: 185–189.

41. Leurs LJ, Schouten LJ, Mons MN *et al.* (2010) Relationship between tap water hardness, magnesium and calcium concentration and mortality due to ischemic heart disease or stroke in the Netherlands. *Environ. Health Perspect.* **118**: 414–420.

42. Sauvant MP, Pepin D. (2002) Drinking water and cardiovascular disease. *Food Chem. Toxicol.* **40**: 1311–1325.

43. Lake IR, Swift L, Catling LA *et al.* (2010) Effect of water hardness on cardiovascular mortality: An ecological time series approach. *J. Public Health (Oxf)* **32**: 479–487.

44. Witlin AG, Sibai BM. (1998) Magnesium sulfate therapy in preeclampsia and eclampsia. *Obstet. Gynecol.* **92**: 883–889.

45. Guerrera MP, Volpe SL, Mao JJ. (2009) Therapeutic uses of magnesium. *Am. Fam. Physician.* **80**: 157–162.

46. Chaumont A, Voisin C, Sardella A, Bernard A. (2012) Interactions between domestic water hardness, infant swimming and atopy in the development of childhood eczema. *Environ. Res.* **116**: 52–57.

# CHAPTER 11

# FUNDAMENTALS AND HISTORY OF RADIATION AND RADIOACTIVITY

The first hurdle the public faced in understanding the impact of radiation on health is the myriad of technical terms that accompany information from news media and the scientific community. A survey examining non-experts' conceptions of radiation showed that the public does not grasp the terms and concepts used in explaining radiation and radioactivity.[1] Therefore, before exploring the hazards of naturally occurring radiation, we should understand some fundamentals about types of radiation and measurement of radioactivity, at least the most frequently encountered terms such as radon; alpha ($\alpha$), beta ($\beta$) and gamma ($\gamma$) radiations; and units such as gray, roentgen, rads, REM and sieverts.

There is a popular tendency to equate radiation with radioactivity. In scientific terms, radioactivity is just one part of radiation. By definition, radioactivity is the property of certain species of atoms and elements (nuclides) to spontaneously transform into another element and in the process emitting radiations such as alpha ($\alpha$), beta ($\beta$) and gamma ($\gamma$). Radioactivity is measured by the number of disintegrations of the nuclides. Radiation, on the other hand, is a term used to describe any and all processes that transmit energy through space or through a material medium. The process of transmission can be in the form of (1) electromagnetic waves or (2) energetic particles.

## Electromagnetic Radiation (EM)

Electromagnetic radiation spectrum spans both the non-ionizing and ionizing radiation as defined by the frequency and wavelength

181

(Fig. 1). Within these two broad groups, radiation emission ranges from low energy (low frequency and high wavelength) radio waves to the high energy (high frequency and low wavelength) γ-ray, with microwave, infrared, visible light, UV light, and X-ray situated in between (Fig. 1). During radioactive decay, some radionuclides such as cobalt-60, zinc-65, cesium-137, and radium-226 emit packets of electromagnetic energy in the form of X-ray and γ-ray (Fig. 1) or "photons".[a] Because they have no mass, X-ray and γ-ray can travel a long distance and can penetrate body tissues to cause damages; they are typically stopped by using shields made of dense materials such as lead.

## Particle Radiation

Some types of radiations are best described in terms of energy particles. The atomic structures of some nuclides and isotopes are unstable and they spontaneously transform or decay to a different, more stable state. During these transformation or decaying processes, excess energy is radiated in the form of electromagnetic wave (EM) and/or particles — α and β particles. An alpha particle consists of two neutrons and two protons and is positively charges. Because of its heavy mass, α particle cannot travel more than a few inches and can be stopped by a sheet of paper. Typical α emitters are uranium and radon. Beta particles, in contrast, are negatively charged electrons emitted from the nucleus of radioactive atoms such as H-3 (tritium), carbon-14 and phosphorus-32. Depending on the energy level it possesses, β particles may travel up to several feet and penetrate the skin. Although α and β particles are much less penetrating than X-ray and γ-ray they are nevertheless hazardous if swallowed or if they are in close contact with the body and skin.

---

[a] Photon is a quantity of electromagnetic energy ($E$) whose values in joules is the product of its frequency ($v$) in hertz times the Planck Constant ($h$): $E = hv$. Thus the higher the frequency, the higher the energy. Photon has no mass or charges like alpha and beta particles.

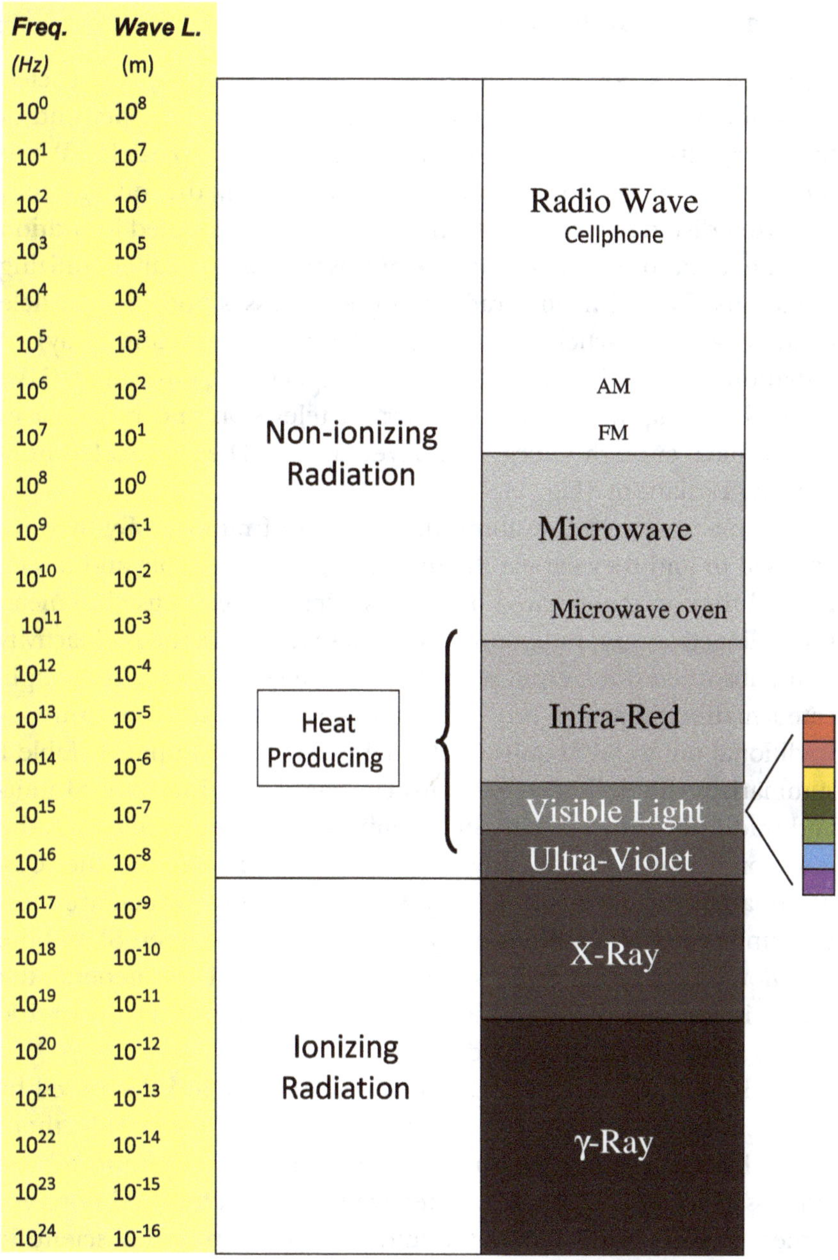

**Figure 1.** Electromagnetic spectrum of ionizing and non-ionizing radiation.

## Ionizing and Non-Ionizing Radiation

Alpha particles possess the highest energy followed by $\beta$ particles. X-ray and $\gamma$ ray, which are situated at the highest frequency end of the EM spectrum, possess lower energy than $\alpha$ and $\beta$ particles. When these types of radiations impinge on an atom, the energy they imparted dislodge electrons from it. The process is called ionization and the radiations that are capable of this process are called ionizing radiations. Thus, ionizing radiation encompasses both the particle form ($\alpha$ and $\beta$ particles) and the EM forms (X-ray and $\gamma$-ray) of radiations. At the lower end of the EM spectrum, ultraviolet (UV) light, visible light, infrared, microwave, television and radio waves do not have enough energy to ionized atoms. They are called non-ionizing radiations (Fig. 1).

People want to know about the amount of radioactivity they are exposed to and they expect the information to be simple and direct. Instead they are confronted by various scientific terms used to measure radioactivity and radiation doses. These terms are the radioactivity units, absorbed dose, equivalent dose, and effective dose. To complicated matter there are two co-existing systems of nomenclature — traditional units and SI units (Système international d'unités). Table 1 summarizes the definition of terms commonly used in both SI units and traditional units. While the SI units are used internationally, the traditional units (curie, gray, rem) are still in use in the U.S. Fortunately, the absorbed dose and equivalent dose, which are most encountered by the public, are easily convertible between SI and the traditional unit with 1 Sv equal to 100 rem (Table 1). In general, the public is exposed to lower doses of radiation and mSv (1 mSv = 100 mrem) is usually the unit used.

It should be noted that the meaning of radiation dose used by the scientific community can be quite confusing. For example, there are at least five different ways to express radon dose, and all are expressed in mSv.[20] The term effective dose and absorbed dose are sometimes used indiscriminately and without definition in scientific literature and public media. In general when only "dose" is mentioned, it refers the equivalent dose.

**Table 1.** Measurement of radioactivity and radiation in SI and traditional units.

| | SI Unit | Traditional unit |
|---|---|---|
| **Radioactivity** | | |
| Definition: The activity of a radioactive substance measured by the number of disintegration of the atom per second | 1 Becquerel (Bq) = 1 disintegration per second | 1 curie (Ci) = 3.7 × $10^{10}$ disintegration per second |
| | $1\ Ci = 3.7 \times 10^{10}\ Bq$ | |
| **Absorbed dose** | | |
| Definition: Amount of ionizing radiation imparted on a mass or body | 1 gray (Gy) = 1 joule/kg | 1 rad = 0.01 joule/kg |
| | 1 Gy =100 rad | |
| **Equivalent dose** | | |
| Definition: The absorbed dose that takes into account the type of radiation being absorbed. It is the absorbed dose multiplied by a radiation weighting factor ($W_R$)* | 1 Sv (sievert) = 1 Gy × $W_R$ | 1 rem (roentgen equivalent man) = 1 rad × $W_R$ |
| | 100 rem = 1 Sv | |
| **Effective dose** | | |
| Definition: Sum of equivalent doses, weighted by the appropriate tissue weighting factor ($W_T$)* | Sum of (effective dose × $W_T$) in Sv | Sum of (effective dose × $W_T$) in rem |
| | 100 rem = 1 Sv | |

*Radiation weighting factor (WR) is 1 for $\gamma$-radiation, X-radiation and $\beta$ particles, 5–20 for neutrons of various energy levels, and 20 for $\alpha$ particles; tissue weighting factor (WT) is 0.2 for testes or ovaries, 0.12 for bone marrow, lung, stomach and colon, 0.05 for liver, breast, bladder, esophagus and thyroid gland, and 0.01 for skin and bone.[21]

## History of Radiation Exposure

Radiation existed at the creation of the universe. It is the most pervasive, constant and penetrating natural energy we are exposed to. The term radiation elicits a mixed emotion from the general public.

We know very well we benefit from the use of radiation in medicine — X-ray, computer assisted tomography, and radiation therapy in cancer treatment, to name a few applications. Our society's huge hunger for energy is satiated in part by power derived from nuclear plants. In some countries such as France, Lithuania and Belgian, nuclear power is the major source of energy. Radioisotopes are indispensible tools in biological research. Radiation is widely used in the manufacturing and construction businesses for material evaluation and gauging, product sterilization and quality control. It is employed in oil exploration for well logging — a method of assessing the porosity of the rock formation. Radiography is used in the oil industry to check for defects and cracks in pipelines. The scanners at the airports (whole body and luggage scanners) also use radiation. Even smoke detectors in our homes are very mildly radioactive.

The public is keenly aware of the negative aspects of radiation. The destructive power unleashed by the atomic bombs in Nagasaki and Hiroshima and the horrible human suffering they wrought is forever etched into the collective memory of mankind. During the Cold War period, the seemingly endless series of nuclear tests jolted our sense of security. Those who lived through the period will recall news announcing nuclear tests that were conducted and, more importantly, the direction the radioactive clouds were heading. Our anxiety towards radiation was exacerbated by a string of nuclear mishaps that occured throughout the world. The first and least publicized event occurred in 1957 in Kyshtym, the former Soviet Union, where a nuclear fuel reprocessing plant exploded. The event was assessed at level 6 on a seven-point event scale by the International Atomic Energy Agency.[22] This was followed by accidents in nuclear generating stations at Three Miles Island, U.S. in 1979 (level-5), Chernobyl, the then Soviet Union in 1980, and Fukushima, Japan in March 2011. The most recent mishap occurred in Marcoule, France in September 2011, where a nuclear waste processing plant exploded. The accidents at Chernobyl and Fukushima were rated seven, the highest in the IAEA event scale.

The era of radiation arrived with the discovery of X-ray by the German scientist Wilhelm Roentgen in 1895, and radioactivity by the French scientist Henri Becquerel in 1896. At the time, no one was aware of the danger of radiation. Otherwise it would be hard to imagine that Wilhelm Roentgen would have allowed his wife to put her hand in the path of the X-ray to develop an image of bones, wedding ring and all.[2] The general public in fact was quite eager to see what kind of benefits radiation may bring.[25] The period between 1900 and 1950 was an enthralling and an innocent time indeed, as illustrated by the following stories.

## The Unsuspecting Era (1900–1950)

*Marie Curie.*[3,4] Madam Curie (1867–1934) was by all measures one of the most esteemed female scientist. She was the first woman Nobel laureate, winning the prizes twice, first for physics in 1903 and then for chemistry in 1911. She discovered the radioactive isotopes polonium and radium, and devoted much her time studying a novel and strange phenomenon: a penetrating ray call radiation. She was isolating naturally radioactive materials in a poorly ventilated garage. Safety measures such as protective clothing, goggles, masks, Geiger counters, and negative ventilation, were unheard of at the time. Radioactive materials were stored in her desk drawers and carried around in her pocket with no shielding materials. She was quoted as saying that she marveled at the luminous glow of the purified elements that looks like fairy lights. It was said that her personal items and stationary are still radioactive today. The exposure took its toll. She suffered radioactive burn on her fingers and hands. She died of aplastic leukemia — a blood disorder in which the body's bone marrow does not make enough new blood cells. A primary cause of the disease was excessive exposure to radiation.

*The shoe-fitting pedoscope.* Industry was quick to exploit the penetrating power of radiation for business purposes. In the mid-1920s, inventors in the U.S. and Britain came up with the idea of shoe-fitting fluoroscope or pedoscope. The contraption consist of an X-ray emitting tube at the bottom and a fluorescent screen on top. Customers would place their feet into a cavity at the bottom of the machine and examine the image of the bones in their feet. This supposedly helps them decide which shoes best fit their feet. Shoe

**Figure 2.** A 1950s children's book depicting a trip to a shoe store and the use of a pedoscope[24] by a child.

stores welcomed the machine as a sales gimmick and customers embraced it as a novelty. Adults and children alike were encouraged to use the machine, which was operated by store clerks with no training (Fig. 2). Nor were there any warnings about potential hazards.

When scientists and medical professions gradually became aware of the danger of X-ray, they also raised concern about the pedoscopes. Some government agencies started to regulate or ban its use. By the late 1950s, very few machines were left although some claimed that they could still be found into the 1970s.[5] At the peak of its use, there were an estimated 10,000 units in use in the U.S., 3000 in Britain and 1000 in Canada.[6] People most affected by the machine appeared to be workers at the shoe stores. Decades after the disappearance of the pedoscopes, reports started to appear in medical journals describing a rare basal cell cancer of the foot or sole.[6,7] These published cases involved two women — 72 and 82 years old, both of them recalled working at shoe stores and frequently operated the pedoscopes.

***Radioactive spas.*** In the 1920s, numerous resorts were built worldwide with the selling point that the water and the air contained trace amount of radium. Some claimed that the radioactivity possessed curative power towards chronic diseases. Others touted the stimulatory and rejuvenating effects of the radioactive elements on the body. With the increased awareness of the hazards of radiation, these spas have gradually disappeared. In Sweden, radioactive spas are no longer allowed.[9] A few radioactive spas still exist in Japan, United Sates and Europe, but now the claims are different. Patrons are promised beneficial effects such as activation of antioxidation functions and alleviation of oxidative damages.[8] Supporters still believe the radiation provides long-term beneficial effects for patients with chronic ailments such as rheumatoid arthritis.[10] A radon spa in Gastein, Austria, has a website promoting "Therapy for pain relief, regeneration, and relaxation". A hospital in Hungary presently has a "medical bath" with therapeutic pools that are fed with water containing radon. G. Trabidou and H. Florou[23] investigated the inhabitants of the island of Ikaria, Greece, and bathers and spa workers of the radioactive spas on the island. They found that the spa workers were exposed to radiation doses of up to 35 mSv/year, which exceeds the recommended limit of 20 mSv/year for workers, recommended by the International Commission on Radiological Protection (ICRP).[11]

***Radithor.*** While early radium therapy in hospitals paved the way for the modern cancer radiotherapy, the "consumer" use of radium was an unmitigated disaster. Radithor was a patent medicine marketed in the U.S. and worldwide in the 1920s. It guaranteed to contain radium in distilled water and claimed to be a tonic for over 150 "endocrinologic diseases", even sexual impotence. The U.S. Government issued a number of warnings against Radithor for "false representations" but the effect on the public was minimal.[25] It took the case of Eben Byers, a millionaire sportsman, to arouse the public to the hazard of radium. Byers took Radithor on the advice of a therapist for over two years. He suffered severe headaches, jaw pains, and "disintegrating bone". At his death, he weighted just 92 pounds.[12,25] We now know he clearly suffered from radium poisoning. It was estimated that the amount of radium in his bones was equivalent to a dose of 350Sv (350,000 mSv). Such a massive dose, if taken at once would have been inneately fatal (see Table 4, Chapter 12).[13]

***Self-luminous watches and radium girls.***[13–15] Radium and zinc sulfide were the key ingredients in luminous paints. When zinc sulfide is bombarded by the radiation from radium, it emits phosphorescence and glows in the dark. The paint is applied onto dials and watches to make them luminous at night. During the First and Second World Wars, self-luminous dials on military instruments and time pieces were indispensible for battle-field operations. Self-luminous watches were popular consumer items from 1930s to 1950s. They gradually disappeared in the industrialized countries but are still sold in many less developed countries years after. Even today self-luminous dials can be purchased in auction houses as war memorabilia and they are still mildly radioactive. Although radium is completely phased out today, it should be noted that other type of radiation-induced luminescence may still be in use, for example, self-luminous warning and exit signs are used in some countries. Tritium, a radioactive gas, is the radiation source instead of radium. The small amount of tritium is hermetically sealed. In industrialized countries, the sale, distribution, and disposal of these tritium source are regulated and risk of exposure is minimal.

Even though consumer use of radium is in the past, we must not forget the workers who were exposed to it and suffered the health consequences. These are the dial painters of the early 1900s, most of them young women. They worked directly with radium-containing paint with no protective garments and no safety instructions. It was not unusual for them to sharpen the tip of the paintbrush with their fingers or even their lips. Radium, which is similar in chemical structure to calcium, deposited in the bones and affected their growth and metabolism. Many became victims of radium-induced diseases — loose teeth, necrosis of the jaws, fragile bones and sinuses. They also have much higher incidence of bone cancer 40–50 years after first exposure.[16] Five of the girls, the "Radium Girls" as the press called them, sued the company, the U.S. Radium Corporation of New Jersey. They won after bitter and protracted litigations. In her book *"Radium Girls, Women and Industrial Health Reform, 1910–1935"*,[15] Claudia Clark detailed the suffering of young women who worked at dial factories in the U.S. Their courageous fight for justice was a turning point in the history of workplace safety and industrial health. One of the important points raised by Clark was this: *"With the knowledge of the health effects of radium and skills to measure these effects held by a small number of scientist, we must consider whether those with knowledge and skills had a greater ethical duty to share them than if such knowledge were widely available"*. Scientists, medical professionals and government regulators often faced the same dilemma today.

***From mountain sickness to indoor radon.*** Decades before the discovery of radon, a 16th century physician Georgius Agricola noted the frequent occurrence of a mysterious "mountain disease" among silver miners in the Erz Mountains straddling present-day Germany and the Czech Republic. Data collected in the 1800s showed that about 75% of the miners in the nearby Schneeberg region died of lung diseases.[9] In 1879, F.H. Harding and W. Hesse identified the miners' malady as pulmonary malignancy. With the expansion of uranium mining following World War II, concerns about miners' health prompted several epidemiological studies which firmly linked radon to the increased rates of lung cancers.[17] In 1966, the Colorado Department of Health discovered that uranium mine tailings had been used as fillers under new buildings, and as aggregate in concrete and masonry building materials. As a result, the indoor radon levels in these building approached or surpassed occupational exposure limits. By 1969, a survey of 250 homes in Grand Junction, Colorado, where uranium mine tailings were used, revealed radon level above the permissible levels.[18] And the era of indoor radon began.

Within 30 years of the discovery of radiation, a body of knowledge had been accumulated on the adverse effects of radium and other radioactive elements. Government and international agencies were set up in response to the concern of the public. In 1928, an independent non-government body of experts in the field, the International X-ray and Radium Protection Committee, was established. It was later renamed the International Commission on Radiological Protection (ICRP).[19]

## References

1. Henrikson EK. (1996) Laypeople's understanding of radioactivity and radiation. *Radiat. Prot. Dosimetry* **68**: 191–196
2. The Nobel Prize in Physics. (1901) Wilhelm Conrad Röntgen. Biography. [http://www.nobelprize.org/nobel_prizes/physics/laureates/1901/rontgen-bio.html].
3. Coppes-Zantinga AR, Coppes MJ. (1998) Madame Marie Curie (1867–1934): A giant connecting two centuries. *Am. J. Roentgenol.* **171**: 1453–1457.
4. Mould RF. (1999) Marie and Pierre Curie and radium: History, mystery and discovery. *Med. Phys.* **26**: 1766–1772.
5. Duffin J, Hayter CR. (2000) Baring the sole. The rise and fall of the shoe-fitting fluoroscope. *Isis.* **91**: 260–282.
6. Smullen MJ, Bertler DE. (2007) Basal cell carcinoma of the sole: Possible association with the shoe-fitting fluoroscope. *Wisconsin Medical J.* **106**: 275–278.
7. Oster-Schmidt C, Altmeyer P, Stücker M. (2002) Basal cell carcinoma of the foot. Is the pedoscope a risk factor? *Hautarzt.* **53**(12): 819–821.
8. Kataoka T, Sakoda A, Yoshimoto M *et al.* (2011) Studies on possibility for alleviation of lifestyle diseases by low-dose irradiation or radon inhalation. *Radiat. Prot. Dosimetry* **146**: 360–363.
9. Swedjemark GA. (2004) The history of radon from a Swedish perspective. *Radiat. Prot. Dosimetry.* **109**: 421–426.
10. Franke A, Reiner L, Resch KL. (2007) Long-term benefit of radon spa therapy in the rehabilitation of rheumatoid arthritis: A randomised, double-blinded trial. *Rheumatol Int.* **27**: 703–713.
11. International Commission on Radiological Protection Publication 103. (2007) *The 2007 Recommendations of the International Commission on Radiological Protection.* Elsevier.

12. Macklis RM, Bellerive MR, Humm JL. (1990) The radiotoxicology of Radithor. Analysis of an early case of iatrogenic poisoning by a radioactive patent medicine. *J. Am. Med. Assoc.* **264**: 619–621.

13. Evans RD. (1933) Radium poisoning. A review of present knowledge. *Am. J. Public Health* **23**: 1017–1023.

14. Boyd JT, Court Brown WM, Vennart J, Woodcocok GE *et al.* (1966) Chromosome studies on women formerly employed as luminous-dial painters. *BMJ* **1**: 377–382.

15. Clark C. (1997) *Radium Girls: Women and Industrial Health Reform, 1910–1935.* University of North Carolina Press.

16. Rundo J, Keane AT, Lucas HF *et al.* (1986) Current (1984) status of the study of 226Ra and 228Ra in humans at the center for Human Radiobiology. *Strahlentherapie* [Sonderb] **80**: 14–21.

17. Kabat GC. (2008) *Hyping Health Risks. Environmental Hazards in Daily Life and the Science of Epidemiology,* Chap. 5. Columbia University Press, New York.

18. Angell WJ. (2008) The US radon problem, policy, program and industry: Achievements, challenges and strategies. *Radiat. Prot. Dosimetry* **130**: 8–13.

19. Clarke RH, Valentin J. (2009) The history of ICRP and the evolution of its policies. *Ann ICRP* **109**: 75–69. ICRP Publication, Elsevier Ltd.

20. Chen J. (2005) A review of radon doses. *Radiat. Prot. Manage.* **22**: 27–31.

21. Canadian Centre for Occupational Health and Safety. (2007) Quantities and Units of Ionizing Radiation. [http://www.ccohs.ca/oshanswers/phys_agents/ionizing.html].

22. INES. The International Nuclear and Radiological Event Scale, International Atomic Energy Agency, Information Series/Division of Public Information. 08–26941/E. [http://www.iaea.org/Publications/Factsheets/English/ines.pdf].

23. Trabidou G, Florou H. (2010) Estimation of dose rates to humans exposed to elevated natural radioactivity through different pathways in the Island if Ikaria, Greece. *Radiat. Prot. Dosim.* **142**: 378–384.

24. *Five in the Family.* Baruch D and Montgomery E. W. J. Gadge and Co. Ltd Toronto.

25. Macklis RM. (1990) Radithor and the era of mild radium therapy. *JAMA.* **264**: 614–618.

# CHAPTER 12

# EFFECT OF IONIZING RADIATION ON HEALTH

*"We must bear in mind that all of us are continuously inhaling the radium and thorium emenations and their products and ionized air. In addition, we are continuously undergoing a type of mild X-ray treatment for the β and γ rays from the Earth and atmosphere that continuously pass into and through our bodies."*

Ernest Rutherford (1871–1937)[1]

## Natural Source of Exposure

The crust of the earth has contained radioactive materials ever since its creation. The earth is also continuously bombarded by high-energy particles which originated in outer space. Throughout lives, humans are exposed to four sources of natural radiation: Cosmic, external/terrestrial, inhalation and ingestion (Table 1).[2,3,74]

### Cosmic radiation

The origin of cosmic radiation is summarized below:

1. Galactic cosmic radiation. Cosmic ray originates from deep space outside the solar system. High energy cosmic particles impinging on the earth's atmosphere interacted with atoms and molecules in the air and generated sets of secondary, less energetic, charged and uncharged particles. At ground level, muons (energetic charged particles similar to electrons but with a heavier mass) are the most important contributor to radiation dose, while neutrons, electrons, positrons, photons and protons

**Table 1.** Estimated Average Annual Dose (mSv) of Natural Radiation Received by Man.

|  | World-Wide[6] | Canada[4] | U.S.[23] |
|---|---|---|---|
| Cosmic/space | 0.39 | 0.3 | 0.3 |
| External/Terrestrial | 0.48 | 0.2 | 0.2 |
| Inhalation (Radon, Thoron) | 1.26 | 0.9 | 2.3 |
| Ingestion/Internal | 0.29 | 0.3 | 0.3 |
| Total | 2.42 | 1.8 | 3.1 |

are the most significant components at aircraft altitude. The earth's geomagnetic field deflects charged cosmic ray towards the magnetic poles, which is the reason why cosmic radiation is more intense near the North and South poles.[74] At time of maximum solar activity, which occurs on an approximately 11-year cycle, the galactic cosmic radiation is at its lowest.[83]

2. Solar cosmic radiation. Magnetic disturbance on the Sun's surface ejects energetic particles, mostly in the form of protons. Some of these particles reach the Earth and contributed to the total intensity of cosmic radiation. During the height of a solar storm, when high concentrations of energetic solar particles reach the Earth, aircrew and passengers may receive a dose of about 1 mSv per subsonic fligh.[9] In other times, however, solar flares do not have sufficient energy to significantly increase the intensity of cosmic radiation at commercial aircraft altitude.[74]

3. Van Allen radiation belt. The Van Allen belts are formed through the capture of protons and electrons by the earth's magnetic field. There are two van Allen radiation belts, an internal one centered about 3,000 km and an external one centered at around 22,000 km above the earth. This radiation is only of concern to space travelers.

***High Altitude Exposure.*** Cosmic radiation is stronger at high altitudes where the atmosphere is thinner. For example, the annual effective dose of radiation from cosmic rays in Vancouver, British Columbia, which is at sea level, is about 0.30 mSv. By comparison a person at the top of Mount Lorne, Yukon (2,000 m) would receive an annual dose of about 0.84 mSv.[4] Similarly, residents of Amargosa Valley, Nevada, and Leadveille, Colorado (altitude 3200 m) are found to have an annual dose rate from cosmic radiation of 0.33 and 0.85 mSv.[5]

*Air Travel.* At commercial aircraft altitudes (7,000–12,000 m for commercial, ~18,000 m for supersonic) the protective layer of the Earth's atmosphere is much thinner than it is on sea level. The intensity of cosmic radiation is approximately 100 times greater at these altitudes than it is on the ground. As a rule of thumb, aircrafts typically travel at 6,100–12,200 m where the dose rate doubles for every 1830 m of increase in altitude. In addition to the altitude, radiation doses received on a flight also depend on the path, duration and latitude of the flight, as well as on the level of solar activity. For a typical one-way flight from Northern Europe to North America, radiation dose received by a passenger is around 4–8μSv/hr or 0.03–0.06 mSv for an 8-hour flight.[6] It was estimated that aircrew received an average annual dose of 2–3 mSv for long-haul and 1–2 mSv for short-haul,[6,7] although estimates as high as 9 mSv have been reported.[8]

### Terrestrial radiation

Terrestrial radiation originates from soils, rocks, water, air and vegetations that contain trace amount of radioactive substances. The main contributor to external exposure comes from gamma-emitting radionuclides such as K-40, U-238 and Th-232 families. Indoor exposure depends on radionuclide concentrations in outdoor soil, building materials and the type of house structure.

*High Background Radiation Area.* On a nation-by-nation basis, the average natural absorbed radiation dose (measured in air) is 55 nGy/hr with a range of 24 and 85 nGy/hr.[10] While average absorbed (equivalent) dose does not differ greatly from country to country, there are certain areas in the world where doses can be exceptionally high. The known high background radiation areas (HBRAs) include Guarapari in Brazil, Yangiang in China, Kerela in India, the Nile delta in Egypt, and the Ramsar and Mahallat regions of Iran. The region of Ramsar in Iran has very high background radiation due to radium-226 deposited from water flowing from hot springs. Residents in the area may receive an effective dose as high as 260 mSv/y due to background radiation,[11,12] which is approximately 110 times the world average of 2.4 mSv (Table 3). In Chinavillai, India, background radioactivity is high due to beach sands that contain natural radionuclides which are a component of the mineral monazite, and maximum ambient dose equivalent as high as 162.7 mSv/y has been recorded (Table 3).[13] It must be emphasized that the definition of effective dose, absorbed dose and ambient dose equivalents, although all in mSv, were not clearly defined in all the studies and may not be interchangeable and comparable. The lack of harmonization in the expression of radiation dose added to the confusion in interpreting scientific information (see Chapter 11, *Ionizing and Non-Ionizing Radiation*).

### Inhalation

Inhaled Radon (Ra-222) is the major contributor to the natural radiation received by the public. Radon is a heavy, colorless and odorless gas emanating from uranium and radium and is naturally present in soil, rocks and building materials. It has a relatively short half-life of 3.8 days and originates from the decay of radium (Ra-226) which in turn is a decay product of uranium (U-238). A minor component of the radioactive gas is thoron (Ra-222) which is the decay product of thorium (Th-232). An important characteristic of radon is that it is presented both outdoor and indoor.

***Indoor Radon.*** Radon gas emanating from rocks and soils and some building materials tends to concentrate in enclosed spaces. More radon from rocks and soils will seep into a dwelling if it is not properly sealed or if there are cracks in the foundation. Radon that enters an enclosed space, such as a family home, an office or a school building can sometimes accumulate to high levels. Buildings that are poorly ventilated will also build up higher levels of radon gas. People who spent more time in these types of structures will be the most at risk. Many countries are active in addressing the indoor radon issue through prevention, mitigation and remedial measures.[3,62,75] These measures include sealing of radon entry routes, improving ventilation, and reversing air pressure differences between the indoor space and the outdoor soil through depressurization techniques.

Effort is being made to map indoor radon levels country-by-country, but progress has been uneven. A radon map covering almost all of Europe has been published.[14,15] As a result, several countries have been identified as having higher than normal levels of indoor radon. These countries are the Czech Republic, Finland, Luxembourg and Sweden, with mean radon levels of 140, 120, 110 and 108 Bq/m$^3$, respectively. In comparison, the world average level of indoor radon is 39 Bq/m$^3$ as estimated by the World Health Organization.[3] Countries such as Austria, Canada, China, India, Spain and the U.S. also have radon maps at various stages of development.[16-20] Some regions of the world have been identified as having very high levels of indoor radon. For example, Villar de la Yegua, located in Salamanca, Spain has indoor radon concentrations of up to 15,000 Bq m$^{-3}$, corresponding to an effective dose of up to 110 mSv/y.[21] Indoor radon level as high as 2,487 Bq/m$^3$ has been measured in Niska Banja region of the Balkans(Table 2).[22] These levels are hundreds of fold higher than the world average.

Geographically- and geologically- based radon maps provide the general public with an idea of the probability of exposure to radon in a certain region or locality. However, radon maps do not have the required resolution to pinpoint local areas of concern.[82] Moreover, indoor radon levels in a building or house are dependent on factors such as characteristics of the building materials and surrounding soils, house design and ventilation. Therefore if a person wishes to know about the risk of indoor radon exposure, measurement of radon level at his/her residence and work area is

**Table 2.** Annual Per Caput Radiation Dose, Global and the United States (mSv).[2]

|  | Global | United States | |
|---|---|---|---|
|  | 1997–2007 | 1980 | 2006 |
| Natural Background | 2.4 | 2.4 | 3.1 |
| Medical-related | 0.62 | 0.53 | 3.0 |
| Others | 0.03 | 0.04 | 0.14 |
| Total | 2.82–3.42 | 2.98 | 6.24 |

**Table 3.** Examples of Estimated Radiation* Dose Received Per Year from Natural Sources.

|  | Sources | Radiation Dose (mSv/y) |
|---|---|---|
| Estimated Average Personal Exposure (annual) | Worldwide average background (cosmic + terrestrial + internal + inhaled)[2] | 2.4 |
|  | Coal mining[2] | 2.4 |
|  | Other mining[2] | 3.0 |
|  | Air crew[2,7,8,84] | 0.3–9 mSv |
|  | Spas, waterworks, tourist caves & visitor mines[2] | 4.8 |
|  | Indoor radon, 300 Bq/m³ (Maximum level)[2] | 10 |
|  | Background radiation (Ramsar, Iran)[12] | Up to 260 |
|  | Background radiation (Chinnavillai, India)[13] | Up to 163 |
|  | Indoor radon (Villa de la Yequa, Spain)[67] | Up to 110 |
|  | Indoor radon (Niska Banja, Balkans)[68] | Up to 63 |

*Radiation dose in this table refers to the equivalent dose. In some of the studies, the term "dose absorbed" was used without definition or specification of the tissue weighing factors. These dose values were taken to mean radiation dose or effective dose (see Chapter 11 *Ionizing and non-ionizing radiation*)).

needed. Many factors affect levels of indoor radon, e.g. ventilation, diurnal and seasonal changes, therefore, it is important that a long-term serial radon measurement be performed to provide a realistic assessment.[3]

### Ingestion/internal radiation

We received internal exposure from radioactive element which entered the body through food and drinks. Radiation doses from ingestion are mainly due to K-40, U-238 and TH-232 series radionuclides naturally present in food and water. The concentrations of naturally occurring radionuclides in food varies widely because of differences in the background levels in soil and water, and the way crops are cultivated and animals are raised. Food consumption customs and practices also contribute to the variation. Ingested uranium and radium are deposited in the bone while K-40 is distributed uniformly throughout the body tissues and fluids (Table 1).

## Total Background Exposure

In addition to natural source of radiation, exposure to man-made radiation originating from medical/diagnostic, consumer and occupational/industrial sources also contribute to total exposure. The total background radiation is summarized in Table 2. It can be seen that medical-related radiations (diagnostic radiology, interventional radiography, nuclear medicine) accounts for a substantial proportion of the total background radiation received.[2] If the trend towards increased medical-related use of radiation continues, it is possible that, in the coming decade, it will exceed natural background radiation as the major contributor to total background radiation dose.

## Health Effects of Naturally Occurring Ionizing Radiation

### Biological effects of ionizing-radiation

Different types of ionizing radiations (cosmic rays, α-radiation, β-radiation, γ-ray and X-ray) possess different amounts of energy.

As they track through mammalian cells, the energy imparted can generate free radicals and knock off electrons within cells, breaking or creating bonds in important biomolecules such as protein and DNA. These actions destabilize and damage the biomolecules, and impair or abolish their ability to perform physiological functions. Cellular responses to ionizing radiation fall into the following categories:

(1) Damages to cell structures, membranes, organelles, functions, etc are so severe that the cells can die. Such damages may occur when the radiation dose is acute and high.
(2) Damage is not severe and the cells recovers.
(3) Damage to the DNA is lethal and the cells die.
(4) Damage to the DNA is repaired and the cell survives.
(5) Damage to DNA is not repaired properly and the abnormal cells divide and grow uncontrollably causing cancer. The damaged DNA may be passed on through subsequent cell divisions, producing mutated cells. These mutations may also contribute to the formation of cancer.
(6) Cells that are directly irradiated may transmit signals to adjacent, but not directly irradiated cells, and cause adverse chromosomal changes and genetic instability to adjacent cells. This is called a "bystander effect".[76]

## Adverse Effects from High Dose Exposure

Normal exposure to naturally occurring ionizing radiation is very low and it is difficult to evaluate its effect. Much of our knowledge about adverse effects was gained from nuclear and radiation events where people were exposed to high doses of radiation. Two of the most well-studied events were the nuclear bomb explosions at Hiroshima and Nagasaki and the Chernobyl nuclear accident. Acute high-dose exposure have also been studied in industrial accidents and medical mishaps.

*Life Span Study.* The study is an undertaking initiated by the United States soon after the Second World War and co-administered with the Japanese government. It follows a group of 120,321 survivors of the atomic blasts at Hiroshima and Nagasaki and monitors their mortality and cause of death.[24–26] Results so far have confirmed that the earliest observed effect of the nuclear blasts and subsequent radiation exposure was leukemia, which was manifested in victims within five years of the explosions. Approximately 10 years after exposure, increase in solid forms of cancers and other forms of cancers over background level were also detected. An important finding was that children were more susceptible to the radiation effect than adults. There were also indications of increased risk of cardiovascular diseases.

***Chernobyl Studies.*** The explosion at the Chernobyl nuclear plant caused the uncontrolled release of enormous amounts of radioactivity into the environment. For almost 10 days, radioactive materials, mostly Iodine-131 and cesium-137, were released and deposited into part of Ukraine, Belarus and the Russian Federation. This event occurred in 1986, 41 year after Hiroshima and Nagasaki, and by that time the world was acutely aware of the danger of high dose radiations. A number of public funded studies were initiated soon after the accident. These studies also benefited from knowledge and expertise accumulated from earlier studies including the life span study. The findings were reviewed by the Chernobyl Forum[27] and summarized in the United Nations Scientific Committee on the Effects of Atomic Radiation (UNSCEAR) 2008 report.[6] It was generally acknowledged that immediately after the accident, 134 plant staffs and emergency workers were exposed to extremely high dose of radiation. Out of these 134 workers, 28 workers died within the first few months; and the main cause of death was bone marrow failure. The report concluded that many people involved in the recovery operation had increased incidence of skin injuries and cataracts, and indication of increased incidence of leukemia. In the exposed population, there was an increased incidence of thyroid cancer, with children and adolescents being the most vulnerable. It was further shown that there was a dose-response relationship with I-131, which was released in large amount from the nuclear explosion.[28,29] UNSCEAR[6] concluded that "*...although those exposed to radioiodine as children or adolescents and the emergency and recovery operation workers who received high doses were at increased risk of radiation-induced effects, the vast majority* *of the population (from the three most affected countries) need not live in fear of serious health consequences from the Chernobyl accident."* The UNSCEAR Report was not met with unanimous acceptance. A 2009 book published by the New York Academy of Sciences was devoted to the Russian language publications on the consequences of Chernobyl.[30] It expressed a number of ecological and health concerns. One of the main criticisms was that UNSCEAR underestimated the dose received and the extent and magnitude of the health impacts. The book also suggested that there were other health consequences such as accelerated aging, brain damage, genetic damage and birth defects. The finding and conclusions of the book were contested by other experts.[31,85,86] Concerns were expressed on methodological approaches and study designs used by various investigators, and on some uncorroborated claims. There is also a suggestion that the greater number of advanced tumors observed shortly after the Chernobyl incident can be explained by the "screening effect" (detection of previously neglected cancers), and that there may be a tendency towards overestimation of the consequences of the incident.[31] One thing that is clear is that the debate on health effects will go on for a long time.

*Acute Exposure.* When a person is acutely exposed to a high dose of ionizing radiation, the immediate symptoms are nausea, fatigue, vomiting and diarrhea. These immediate responses are followed in time by loss of hair, hemorrhage, inflammation of the mouth and throat, and general loss of energy. Depending on the dose received, the patient may die immediately or after a short period of time. In general, 20,000 mSv is immediately fatal while 5000 mSv may cause death within weeks 50% of the time (Table 4).[32] At lower doses, adverse effect ranges from hemorrhage and diarrhea at 900–1000 mSv, to hair loss, vomiting, fatigue at 500–750 mSv. At 50–100 mSv, subtle biochemical effects are detectable in the form of hematological changes (Table 4). For comparison purpose, the doses received by one-time scanning/medical events are summarized in Table 4B

## Carcinogenic Classification of Ionizing Radiations

While debate continues about the extent of cancer risk and types of cancers caused by radiation exposure, there is almost universal

**Table 4.** (A) Acute/accidental Radiation Exposure and Associated Health Effects; (B) One-time Scanning/medical event.[2,32,69-73]

| A Acute/Accidental Exposure | Effective dose (mSv) | Health effect | Time to onset |
|---|---|---|---|
| | ~20,000 | Death, damage of central nervous system | Minutes |
| | ~10,000 | Loss of consciousness | |
| | | Death | Hours to days |
| | 5000 | Destruction of intestinal lining | |
| | | Internal bleeding | |
| | ~4,000 | Death (50% of the time) | 1-2 Weeks |
| | 900-1000 | Possible death | Within 2 months |
| | ~750 | Hemorrhage, diarrhea | |
| | ~700 | Hair loss | 2-3 Weeks |
| | 500-550 | Vomiting | |
| | 50-100 | Fatigue, nausea | |
| | 100 | Changes in blood chemistry | |
| | | 1 cancer case/100 person exposed | Life time |

| B One Time Scanning/Medical Event | Effective dose (mSv) | Event |
|---|---|---|
| | ~20 | Neonatal abdominal CT-scan |
| | ~15 | Barium enema |
| | 5-30 | CT-scan (depending on organ scanned) |
| | ~3.0 | Screening mammography |
| | 0.03-0.06 | Air travel, Europe to N. America, one way |
| | 0.04-0.1 | Chest X-ray |
| | <0.0001 | Backscattered X-ray scanner — airport |

acceptance that ionizing radiations are carcinogens. The following naturally occurring radioactivity and decay products have been classified by the International Agency for Research on Cancer as Group 1, or known human carcinogen.[32]

Radon-222 and its decay products.
X- radiation and Gamma-radiation.
Radium-224 and its decay products.
Radium-226 and its decay products.
Radium-228 and its decay products.
Thorium 232 and its decay products.
Plutonium-239.
Phosphorus-32.
Radioiodines.
Internalized radionuclides that emit α-particles.
Internalized radionuclides that emit β-particles.

The U.S. National Toxicological Program[33] Report on Carcinogen classified the following ionizing radiations as "known to be human carcinogens":

X-radiation.
Gamma radiation.
Neutron.
Radon.
Thorium dioxide.

***Extrapolation from High Doses.*** Although radiation at high doses clearly increases cancer risk, unequivocal evidence of increased cancer risk at low doses is much more difficult to obtain. Extrapolation of risk estimates based on observations at moderate to high doses continues to be the primary basis for estimation of radiation-related risk at low doses and dose rates.[34,77] To estimate the number of cancers produced at low dose levels, agencies such as the International Commission on Radiological Protection (ICRP)[77] and the U.S. National Research Council's Committee on the Biological Effects of Ionizing Radiations (BEIR) adopted an extrapolation methodology

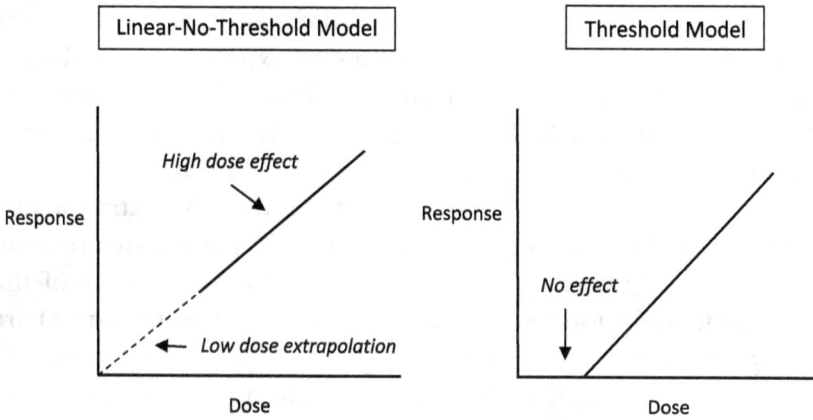

**Figure 1.** Linear-non-threshold (LNT) model for cancer effects and threshold model for non-cancer effects.

based on a "linear-no-threshold" (LNT) risk model.[35] The LNT model is a dose-response model which is built on the assumption that, in the low dose range, radiation dose greater than zero will increase the rate for excess cancer and/or inheritable disease in a simple, proportional manner (Fig. 1)[a]. Such a model is consistent with the precautionary principle embraced by most risk assessment agencies. For radiation protection agencies, the principle is reflected in the belief that the level of radiation the general public is exposed to should be "As Low As Reasonably Achievable" (ALARA).[36] It should be noted that for non-cancer effects of radiation, a threshold rather than non-threshold model is used. The threshold model stipulated that there is a dose value below which no observable effects are detected (Fig. 1).

According to the BEIR VII[35] estimation using the LNT model, approximately 1 in 100 persons would be expected to develop cancer in a lifetime, from a dose of 100 mSv. To put in perspective, at the same time, 42 out of the 100 individuals would be expected to

---

[a]In addition to the LNT model, a benchmark dose (BMD) model is now being applied to assist in the determination of regulatory limit for some chemical carcinogens such as arsenic (see Appendix *b* in Chapter 2).

develop cancers from other causes. In other words, only 1 out of the 43 people who developed cancer is due to exposure to a radiation dose of 100 mSv. The US Environmental Protection Agency arrived at a similar if slightly lower estimate of lifetime risk of 0.84 person developing cancer due to exposure to a 100 mSv dose.[37]

A major criticism of the LNT model is that the extrapolation curve at low dose may not be linear and that any departure from linearity would affect the extrapolated results. Proponents of the hormesis model (non-linear or monotonic dose-response curve) are among those who persistently challenged the LNT model.[78,79] However, the hormesis model has not received much support from the regulatory community.[80,81]

## Adverse Effects from Low Dose Exposure

*Health Effects in High Background Radiation Areas (HBRAs).* The LNT risk model stipulates that any radiation higher than zero produces a finite and proportional risk for cancer. It follows that all people exposed to normal background radiation are at risk of developing cancer, even if the risk is very low. Residents of HBRAs chronically receive higher than normal levels of natural radiation should therefore be at even higher risk of developing cancer. In very high radiation areas, radiation doses in terms of mSv per year can be as high as 68 time global average (Table 3). It appeared that UNSCEAR[10] had initially hoped that studying the health conditions of residents in HBRAs would yield valuable information. However, the results so far have been less than revealing. No increase in cancer mortality rate was observed in Yanjiang, a HBRA of China.[38] R.R. Nair *et al.*[39] followed a cohort of 69,958 subjects for 10.5 years in Kerela, India and found no excess cancer risk from exposure to terrestrial gamma radiation. A case control study conducted in the same area also did not detect an increase in risk of mental retardation or cleft lip/palate among offspring of parents residing in the high radiation area.[40] The only effect consistently observed in inhabitants of HBRAs in China, Brazil and India was unstable chromosome aberration in peripheral blood lymphocytes.[41–43] It should be noted, however, that some of the studies that reported chromosomal changes also reported no excessive risk of cancer.[44] Thus, after years of study, there is still no concrete evidence that residents of HBRAs suffered excessive rate of cancer or non-cancer diseases. Difficulties in HBRA research is best summarized by the following rather guarded statement from the International Atomic Energy Agency.[45]

*"There are many high natural background radiation areas around the world where the annual radiation dose received by members of the general public is several times higher than the ICRP dose limit for radiation workers. The numbers of people exposed are too small to expect to detect any increases in health effects epidemiologically. Still the fact that there is no evidence so far of any increase does not mean the risk is being totally disregarded......However, at low doses of radiation, there is still considerable uncertainty about the overall effects. It is presumed that exposure to radiation, even at the levels of natural background, may involve some additional risk of cancer. However, this has yet to be established. To determine precisely the risk at low doses by epidemiology would mean observing millions of people at higher and lower dose levels. Such an analysis would be complicated by the absence of a control group which had not been exposed to any radiation. In addition, there are thousands of substances in our everyday life besides radiation that can also cause cancer, including tobacco smoke, ultraviolet light, asbestos, some chemical dyes, fungal toxins in food, viruses, and even heat. Only in exceptional cases is it possible to identify conclusively the cause of a particular cancer."*

*Air Travel.* As shown in Table 3, aircrew may receive a maximum annual radiation dose of between 3 and 9 mSv. A survey conducted on flight attendants estimated that the annual effective dose from all sources of cosmic radiation exposure was 2.5 +/−1.0 mSv, with a mean career dose of 30 mSv. These annual doses are below the limit of 20 mSv per year recommended for occupational exposure.[46] It should be noted that for pregnant women, the recommended limit for occupational exposure was only 2 mSv. A study conducted by the U.S. National Institute for Occupational Safety and Health founded that airline pilots with long-term flying experience had increased frequency of chromosome translocation.[47] The authors concluded that the pilots had received biologically significant doses of ionizing radiation, but that further well-designed epidemiological studies were needed to clarify the relationship between cosmic radiation exposure and cancer risk. Indeed, recent reviews of epidemiological studies indicated that carcinogenic and other health effects of cosmic radiation on aircrew are still unclear.[48–50] In addition, an ongoing study of German aircrew so far has not seen an increase in general cancer mortality.[51]

Almost two billion people travel by air each year.[48] Not much information is available on the health effects of cosmic radiation on passengers. It can be assumed that because they receive much lower doses of cosmic radiation than aircrew (Table 3), their health risk is proportionally lower. Presently, concern about adverse effects of long-haul flights is focused on venous thromboembolism,[48] which is independent of radiation.

***Indoor Radon and Lung Cancer.*** Radon is the major contributor to the natural ionizing radiation dose received by the general population. For several decades, studies have focused on underground miners who were exposed to high radon concentrations in the work environment. In the early 1980s, knowledge of indoor radon concentrations together with risk estimates based on the studies of mine workers provide evidence that indoor radon may be an important risk factor for lung cancer in the general population (see *From mountain sickness to indoor radon,* Chapter 11).[3,52] Many epidemiological studies have since been conducted to investigate the relationship between indoor radon and lung cancer. Individually these studies are generally too small to provide conclusive evidence, but when analyzed together they provided strong indication that radon is associated with a high risk for lung cancer in the general population.[53–56] The results of the meta-analyses are consistent with a linear dose-response relationship with no threshold. Important points related to radon and lung cancer risk are summarized below:

- Classified as a human carcinogen.[33,57]
- Second leading cause of lung cancer.[3]
- 3–14% of lung cancer attributable to radon (the rest is due to smoking).[3]

- An estimated 21,000 lung cancer deaths in the U.S. annually is radon related.[58]
- Lung cancer risk increases by 8–15% per 100 Bq/m³ increase in radon.[3,52,59]
- Risk of radon-related lung cancer is much higher in continuing smokers than non-smokers, with former smokers having an intermediary risk.[60,61]
- Chronic exposure to even very low levels of radon carries a risk of lung cancer.[3,52]
- US EPA guideline value[62] for indoor radon: 148 Bq/m³.
- WHO[3] proposed reference level for indoor radon: 100 Bq/m³.

The position that remedial action is required because indoor radon is a significant risk factor for lung cancer has been endorsed by many regulatory agencies and public health institutions. However there are dissenting opinions. Geoffrey Kabat[63] provided a comprehensive critique of this majority opinion. Among the reservations expressed are (1) the paucity and robustness of data (mostly from miner studies) used to construct the high to low dose extrapolation, (2) many miners were smokers, and the conditions to which the miners were exposed to were different from that of home dwellers, and (3) the fact that smoking causes the overwhelming majority of lung cancers were down played. There are also studies that suggest that indoor radon reduction efforts may not be cost-effective[64] and that cessation of smoking is more cost-effective than reducing radon levels at home.[65]

## Practical Points

For the general population concerned about environmental (non-occupational) exposure to radiation, the following points may be helpful:

- Current (ICRP) advice is that a person should not receive more than 1 mSv above background levels (natural and man-made).

- For the majority of people, exposure to natural sources of radiation remains fairly stable from year to year.
- Radiation from medical and diagnostic procedures accounts for most of the increase in total background radiation received by an individual.
- Increased exposure to natural sources of radiation may occur if an individual

  — travels frequently by air.
  — moves to a high background radiation area.
  — lives in a house or work in a building with high indoor radon.

- Indoor radon is associated with increased risk in lung cancer.
- Indoor radon levels can be lowered by preventive and mitigation measures.
- Infants, children and pregnant women are more vulnerable to effect of radiation.
- Potassium iodide has no protective effect against natural radiation. It is only effective against exposure to I-131.[66]

## References

1. Shaviv G. (2009) *The Life of Stars: The Controversial Inception and Emergence of the Theory of Stellar Structure.* Hebrew University Magnes Press and Springer Verlag, Berlin, Heidelberg.
2. UNSCEAR (2008). United Nations Scientific Committee on the Effects of Ionizing Radiation. 2008 Report to the General Assembly with Scientific Annexes. Volume I. Sources. Scientific Annexes B. Exposures of the public and workers from various sources of radiation United Nations. New York 2010.
3. WHO. (2009) *WHO handbook on Indoor Radon: A public Health Perspective.* World Health Organization.
4. Grasty RL, LaMarre JR. (2004) The Annual effective dose from natural sources of ionizing radiation in Canada. *Radiat. Prot. Dosim.* **108**: 215–226.

5. Moeller DW, Sun LS. (2006) Comparison of natural background dose rates for residents of the Amargosa Valley, NV, to those in Leadville, CO, and the states of Colorado and Nevada. *Health Phys.* **91**: 338–353.

6. UNSCEAR (2008). United Nations Scientific Committee on the Effects of Ionizing Radiation. 2008. Report to the General Assembly with Scientific Annexes. Volume II. Effects. Scientific Annexes C, D, and E. United Nations. New York 2011.

7. Bagshaw M. (2008) Cosmic radiation in commercial aviation. *Travel Med. Infect. Dis.* **6**: 125–127.

8. O'Brien K, Friedberg W, Duke FE *et al.* (1992) The exposure of aircraft crew to radiation of extra-terrestrial origin. *Radiat. Protect. Dosim.* **45**: 145–162.

9. Beck P. (2009) Overview of research on aircraft crew dosimetry during the last solar cycle. *Radiat. Prot. Dosim.* **136**: 244–250.

10. UNSCEAR (1988). United Nations Scientific Committee on the Effects of Ionizing Radiation. 1988 Report. Sources, Effects and Risks of Ionizing Radiation. Appendix A. Exposure from natural sources of radiation.

11. UNSCEAR (2000). United Nations Scientific Committee on the Effects of Ionizing Radiation. 1988 Report Vol. 1. Sources and Effects of Ionizing Radiation. Appendix B. Exposure from natural radiation sources.

12. Ghiassi-nejad M, Mortazavi SM, Cameron JR *et al.* (2002) Very high background radiation areas of Ramsar, Iran: Preliminary biological studies. *Health Phys.* **82**: 87–93.

13. Matsuda N, Brahmanandhan GM, Yoshida M *et al.* (2011) Background radiation and individual dosimetry in the costal area of Tamil Nadu, India. *Radiat. Prot. Dosim.* **146**(1–3): 314–317.

14. Dubois G, Bossew P, Tollefsen T, De Cort M *et al.* (2010) First steps towards a European atlas of natural radiation: Status of the European indoor radon map. *J. Environ. Radioact.* **101**: 786–798.

15. Tollefsen T, Gruber V, Bossew P, De Cort M *et al.* (2011) Status of the European indoor radon map. *Radiat. Prot. Dosim.* **145**: 110–116.

16. Friedmann H, Gröller J. (2010) An approach to improve the Austrian radon potential map by Bayesian statistics. *J. Environ. Radioact.* **101**: 804–808.

17. Chen J, Jiang H, Tracy BL, Zielinski JM *et al.* (2008) A preliminary radon map for Canada according to health region. *Radiat. Prot. Dosim.* **130**: 92–94.

18. Ramachandan TV, Sathish LA. (2011) Nationwide indoor 222Rn and 220Rn map for India: A review. *J. Environ. Radioact.* **102**: 975–986.

19. Poncela LSQ, Fernandez PL, Arozamena JG. (2004) Natural gamma radiation map (MARNA) and indoor radon levels in Spain. *Environ. Int.* **29**: 1091–1096.

20. US EPA Map of Radon zones. [http://www.epa.gov/radon/pdfs/zonemapcolor.pdf].

21. Sáinz C, Quindós LS, Fernández PL *et al.* (2007) High background radiation areas: The case of Villar de la Yegua village (Spain). *Radiat. Prot. Dosim.* **125**: 565–567.

22. Zunić ZS, Celiković I, Tokonami S *et al.* (2010) Collaborative investigations on thoron and radon in some rural communities of Balkans. *Radiat. Prot. Dosim.* **141**: 346–350.

23. NCRP Report No. 160, *Ionizing Radiation Exposure of the Population of the United States*, National Council on Radiation Protection and Measurement, Bethesda, Maryland.

24. Ozasa K, Shimizu Y, Suyama A *et al.* (2011) Studies of the mortality of atomic bomb survivors, report 14, 1950–2003: An overview of cancer and noncancer diseases. *Radiat. Res.*, in press.

25. Shimizu Y, Kodama K, Nishi N *et al.* (2010) Radiation exposure and circulatory disease risk: Hiroshima and Nagasaki atomic bomb survivor data, 1950–2003. *BMJ* **340**, in press.

26. Richardson D, Sugiyama H, Nishi N *et al.* (2009) Lonizing radiation and leukemia mortality among Japanese atomic bomb survivors, 1950–2000. *Radiat. Res.* **172**: 368–382.

27. Chernobyl. Looking back to go forward. *Proc Int Conf.* Vienna, 6–7 September 2005. IAEA. [http://www-pub.iaea.org/MTCD/publications/PDF/ Pub1312_web.pdf].

28. Brenner AV, Tronko MD, Hatch M *et al.* (2011) I-131 Dose response for incident thyroid cancers in Ukraine related to the Chernobyl accident. *Environ. Health Persp.* **119**: 933–939.

29. Baverstock K, Williams D. (2006) The Chernobyl accident 20 years on: An assessment of the health consequences and the international response. *Environ. Health Persp.* **114**: 1312–1317.

30. *Annals of the New York Academy of Sciences.* (2009) Vol. 1181, Chernobyl Consequences of the Catastrophe for People and the Environment, pp. 287–327, November.
31. Jargin SV. (2011) Validity of thyroid cancer incident data following the Chernobyl accident. *Health Phys.* **101**: 754–757.
32. International Agency for Research on Cancer. (2001) *Ionizing Radiation, Part 1: X- and Gamma (g)-Radiation, and Neutrons*, Vol. 78. IARC Press, Lyon, France.
33. U.S. Department of Health and Human Services. Public Health Service. National Toxicology Program. (2011) *Report on Carcinogens*, 12th ed. Report.
34. Tubiana M, Aurengo A, Averbeck D *et al.* (2005) *Dose-Effect Relationships and Estimation of the Carcinogenic Effects of Low Doses of Ionizing Radiation.* Académie Nationale de Médecine, Institut de France-Académie des Sciences, Paris. Report.
35. BEIR VII. (2006) *Biological Effects of Ionizing Radiation VII. Phase 2–Committee to assess health risks from exposure to low levels of ionizing radiation.* National Academy of Sciences, Washington DC; National Academies Press.
36. ICRP. (2007) International Commission on Radiological Protection Publication 103. *The 2007 Recommendations of the International Commission on Radiological Protection.* Elsevier.
37. US EPA. (1999) *Cancer Risk Coefficients for Environmental Exposure to Radio-nuclides.* Federal Guidance Report No. 13. Office of Radiation and Indoor Air. U.S. Environmental Protection Agency. Washington D.C. 20460.
38. Tao Z, Akiba S, Zha Y *et al.* (2012) Cancer and non-cancer mortality among Inhabitants in the high background radiation area of Yangjiang, China (1979–1998). *Health Phys.* **102**: 173–181.
39. Nair RR, Rajan B, Akiba S *et al.* (2009) Background radiation and cancer incidence in Kerala, India-Karanagappally cohort study. *Health Phys.* **96**: 55–66.
40. Koya PK, Chougaonkar MP, Predeep P *et al.* (2010) Effect of low and chronic radiation exposure: A case-control study of mental retardation and cleft lip/palate in the monazite-bearing coastal areas of southern Kerala. *Radiat Res.* **177**: 109–116.
41. BEIR V. (1990) *Biological Effects of Ionizing Radiation V.* National Academic Press: Committee on the Biological Effects of Ionizing Radiation, Washington.

42. Cheriyan VD, Kurien CJ, Das B *et al.* (1999) Genetic monitoring of the human population from high-level natural radiation areas of Kerala on the southwest coast of India. II. Incidence of numerical and structural chromosomal aberrations in the lymphocytes of newborns. *Radiat. Res.* **152**(Suppl. 6): S154–S158.

43. Zakeri F, Rajabpour MR, Haeri SA *et al.* (2011) Chromosome aberrations in peripheral blood lymphocytes of individuals living in high background radiation areas of Ramsar, Iran. *Radiat. Environ. Biophys.* **50**: 571–578.

44. Hendry JH, Simon SL, Wojcik A *et al.* (2009) Human exposure to high natural background radiation: What can it teach us about radiation risks? *J. Radiol. Prot.* **29**: A29–A42.

45. International Atomic Energy Agency. Factsheets & FAQs. [www.iaea. org/Publications/Factsheets/English/radlife.html].

46. ICRP. (1991) International Commission on Radiological Protection Publication 60. *Recommendations of the International Commission on Radiological Protection.* Pergammon Press, Oxford.

47. Yong LC, Sigurdson AJ, Ward EM *et al.* (2009) Increased frequency of chromosome translocation in airline pilots with long-term flying experience. *Occup. Environ. Med.* **66**: 56–62.

48. Silverman D, Gendreau M. (2009) Medical issues associated with commercial flights. *Lancet* **373**: 2067–2077.

49. Kim JN, Lee BM. (2007) Risk factors, health risks and risk management for aircraft personnel and frequent flyers. *J. Toxicol. Environ. Health Part B* **10**: 223–234.

50. Sigurdson AJ, Ron E. (2004) Cosmic radiation exposure and cancer risk among flight crew. *Cancer Invest.* **22**: 743–761.

51. Zeeb H, Hammer GP, Langner I *et al.* (2010) Cancer mortality among German aircrew: Second follow-up. *Radiat. Environ. Biophys.* **49**: 187–194.

52. ICRP. (2010) International Commission on Radiologial Protection Publication 115. Ling cancer risk from radon and progeny and statement on radon. *Annals ICRP* **40**(1).

53. Lubin JH, Wang ZY, Boice JD Jr *et al.* (2004) Risk of lung cancer and residential radon in China: Pooled results of two studies. *Int. J. Cancer* **109**: 132–137.

54. Lubin JH. (2011) Environmental factors in cancer: Radon. *Rev. Environ. Health* **225**: 33–38.

55. Darby S, Hill D, Auvinen A, Barros-Dios JM *et al.* (2004) Radon in homes and risk of lung cancer: Collaborative analysis of individual data from 13 European case-control studies. *BMJ* **330**: 223–227.

56. Darby S, Hill D, Deo H *et al.* (2006) Residential radon and lung can-cer–detailed results of a collaborative analysis of individual data on 7148 persons with lung cancer and 14, 208 persons without lung can-cer from 13 epidemiologic studies in Europe. *Scand. J. Work Environ. Health.* **32**(Suppl 1): 1–83.

57. IARC (1998). International Agency for Research on Cancer. *Man-made mineral Fibres and Radon,* Vol. 43. Lyon, France.

58. US EPA. Radon (Rn). Health Risks. [http://www.epa.gov/radon/heal-thrisks.html].

59. Krewski D, Lubin JH, Zielinski JM *et al.* (2006) A combined analysis of North American case-control studies of residential radon and lung cancer. *J. Toxicol. Environ. Health A* **69**: 533–597.

60. Darby S, Hill D, Auvinen A, Barros-Dios JM *et al.* (2005) Radon in homes and risk of lung cancer: Collaborative analysis of individual data from 13 European case-control studies. *BMJ* **330**: 223.

61. Gray A, Read S, McGale P, Darby S *et al.* (2009) Lung cancer deaths from indoor radon and the cost effectiveness and potential of policies to reduce them. *BMJ* **338**: a3110.

62. U.S. Environmental Protection Agency. (1986a) *A Citizen's Guide to Radon.* U.S. Government Printing Office, EPA-86–004, Washington DC.

63. Karat, Geoffrey C. (2008) *Hyping Health Risks. Environmental Hazards in Daily Life and the Science of Epidemiology.* Chap. 5. Columbia University Press, New York.

64. Petersen ML, Larsen T. (2006) Cost-benefit analyses of radon mitiga-tion projects. *J. Environ. Manage.* **81**: 19–26.

65. Groves-Kirby CJ, Timpson K, Shield G *et al.* (2011) Lung-cancer reduc-tion from smoking cessation and radon remediation: A preliminary cost-analysis in Northampton shire, UK. *Environ. Int.* **37**: 375–382.

66. Vista B. (2003) Experts advice on potassium iodide use. No protection against Dirty Bombs. *J. Am. Med. Assoc.* **289**: 2058.

67. Saenz C, Quindío's LS, Fernandez PL *et al.* (2007) High background radiation areas: The case of Villard de la Yoga village (Spain). *Radiate. Prot. Dosim.* **125**: 565–567.

68. Zunić ZS, Celiković I, Tokonami S *et al.* (2010) Collaborative investigations on thoron and radon in some rural communities of Balkans. *Radiat. Prot. Dosim.* **141**: 346–350.

69. US. EPA. [www.epa.gov/radiation/understand/health_effects.html].

70. Mehta P, Smith-Bindman R. (2011) Airport full-body screening. *Arch. Intern. Med.* **171**: 1112–1115.

71. Brenner DJ, Hall EJ. (2007) Computer tomography–an increasing source of radiation exposure. *New Engl. J. Med.* **357**: 2277–2284.

72. Limaye MR, Severance H. (2011) Pandora's boxes: Questions unleashed in airport scanner debate. *J. Am. Osteopath. Assoc.* **111**: 87–88.

73. Health Canada. (2011) Exposure to Ionizing Radiation Fact Sheet. [http://www.hc-sc.gc.ca/hc-ps/ed-ud/respond/nuclea/expos-eng.php].

74. Health Canada. Environmental and Workplace Health, Cosmic Radiation Exposure and Air Travel. [http://www.hc-sc.gc.ca/ewh-semt/radiation/comsic-cosmique-eng.php].

75. Department for Environment, Food and Rural Affairs, UK. (2005) Radon: A guide to reducing levels in your home. [www.defra.gov.uk].

76. Blyth BJ, Sykes PJ. (2011) Radiation-induced bystander effects: What are they and how relevant are they to human radiation exposures? *Radiat. Res.* **176**: 139–57.

77. ICRP. (2005) International Commission on Radiological Protection Publication 99. Low-dose extrapolation of radiation-related cancer risk. *Ann. ICRP* **35**(4).

78. Cook R, Calabrese EJ. (2006) The importance of hormesis to public health. *Environ. Health Persp.* **114**: 1631–1635.

79. Vaiserman AM. (2010) Radiation hormesis: Historical perspective and implications for low-dose cancer risk assessment. *Dose Response* **8**: 172–191.

80. Thayer KA, Melnick R, Burns K *et al.* (2005) Fundamental flaws of hormesis for public health decisions. *Environ. Health Persp.* **113**: 1271–1276.

81. Little MP, Wakeford R, Tawn EJ *et al.* (2009) Risks associated with low doses and low dose rates of ionizing radiation: Why linearity may be (almost) the best we can do. *Radiology* **251**: 6–12.

82. Friis L, Carter N, Nordman O *et al.* (1999) Validation of a geologically based radon risk map: Are the indoor radon concentrations higher in high-risk areas? *Health Phys.* **77**: 541–544.

83. Singh AK, Singh D, Singh RP. (2011) Impact of golactic cosmic rays on earth's atmosphere and human health. *Atin Enuroic* **45**: 3806–3818.

84. Shea MA, Smart DF. (2000) Cosmic ray implications for human health. *Space Sci Rev.* **93**: 187–205.

85. Jackson D. (2011) Book Review "Chernobyl: consequences of the catastrophe for people and the environment (Annals of the New York Academy of Sciences)" *J. Radiol. Prot.* **31**: 163–166.

86. Balonov MI. (2012) On protecting the inexperienced reader from Chernobyl myths. *J. Radiol. Prot.* **32**: 181–189.

# CHAPTER 13

# ULTRAVIOLET RADIATION

## Source of Ultraviolet Radiation

The sun is the primary source of natural ultraviolet (UV) radiation reaching the Earth. UV radiation occupies a non-ionizing region of the EM spectrum (see Fig. 1, Chapter 11) and is categorized into UVA (315–400 nm), UVB (280–315 nm) and UVC (100–280 nm). The earth's atmosphere acts as an effective UV blocker. By the time solar UV radiation reaches the surface of the earth, it is predominantly in the form of UVA with less than 5% UVB and very little UVC.

The intensity of UV irradiation varies with the time of day, altitude, latitude and the seasons. On Earth, UV irradiation is the strongest near the equator and decreases towards the poles (Fig. 1). Seasonally, UV intensity is stronger in the summer than in the winter. On a daily basis, UV intensity peaks around noon time. Ultraviolet radiation is more intense at higher altitude as the radiation from the sun is less attenuated by the thinner atmosphere. Reflection also contributes significantly to UV irradiation. A grass lawn scatters 2–5% of incident UV radiation. Dry beach sand reflects about 10–25%. Fresh snow may reflect up to 85–90% of incident UV while water, in particular white foams in the sea, may reflect up to 30%. Clouds influence UV ground irradiance, through reflection, refraction, absorption and scattering, and may increase or decrease UV ground irradiance.[21] A factor leading to increased UV reaching the earth is the depletion of the stratospheric ozone

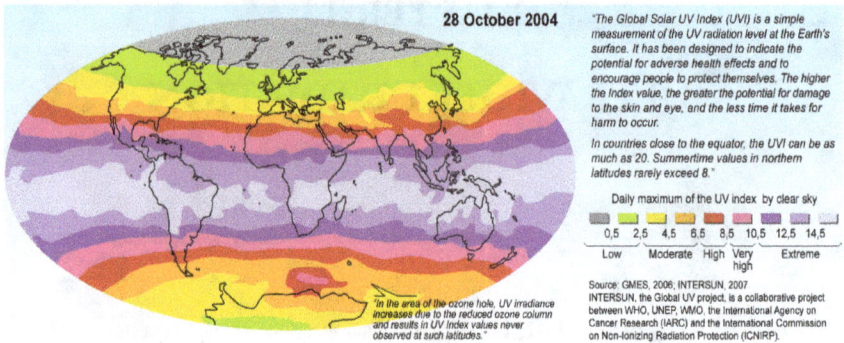

**Figure 1.**   The Global Solar UV Index. 2007. (UNEP GRID Aridal)

*Source*: GMES, 2006; INTERSUN, 2007. INTERSUN, the Global UV project, is a collaborative project between the World Health Organization, United Nations Environmental Programme, World Meteorological Organization, the International Agency for Research on Cancer, and the International Commission on Non-Ionizing Radiation Protection.

layer[a] by halogen ions originated from man-made chemicals. An ozone "hole" is a dynamic region in the stratosphere over the Antarctica (Fig. 1) and the Arctic, where the ozone concentration is 50% or less of normal.[2–4]

### History

The sun must have occupied an exceptional position in the conscious mind of our early ancestors as it provided light, warmth, and security from darkness. The crops they cultivated also depended on the sun. Psalmists in the Old Testaments praise God as the one who makes the sun and directs its path.[5] Hieroglyphic-, cuneiform- and alphabet-based writings between 2000 and 500 BC indicated that the sun was revered as a god by Egyptians, Assyrians, Persians and Babylonians. The sun was also deified by other ancient civilizations

---

[a]The ozone layer is located in the upper part of the earth's atmosphere (stratosphere). Ozone molecules are formed by the action of the sun's UV radiation on oxygen molecules. The ozone layer absorbs the most dangerous UVC and much of the UVB radiations. Ozone is destroyed by the catalytic action of halogen ions originating from man-made chemicals such as chlorofluorocarbons (CFCs) and hydrochlorofluorocarbons (HCFCs).

including the Druids, Aztecs and Incas. Ancient Greek worshipped the sun god Helios.[6]

Approximately 2000 years ago, Inuit (Eskimos) in the Arctic and native Siberians already wore protective eye wears. They are fashioned from tusks or wood in which a horizontal slit was cut out. They effectively allowed peripheral vision while blocking out lights reflected by snow and ice. Although documented cases are lacking, eye black, a form of face paint applied to the cheekbone, was probably used by primitive hunters and warriors to improve their vision under the glare of the sun. Professional baseball and football players today also applied anti-glares on their cheekbone. A scientific study showed that these products actually improves contrast sensitivity when athletes play under the sun.[7]

*Heliotherapy (sunbathing).* Early recognition of the health benefits of the sun's ray may be traced back to around 500 BC when Hippocrates prescribed heliotherapy (sunbathing) for medical and psychological purposes.[6] According to Philip Hockberger,[6] the practice of heliotherapy continued through the Greco-Roman era until the fall of the Roman Empire. Heliotherapy regained its popularity during the early Middle Ages. The Persian scholar and physician Avicenna (980–1037 AD) wrote extensively about the benefits of sunbathing. Sunbathing for medical and cosmetic purposes has continued to the present time with periodic ebbs and tides.[8] The modern, scientific, version of heliotherapy begins with the work of the Danish physician, Niels Finsen (1860–1904), who reported the successful clinical use of ultraviolet radiation in the treatment of tuberculosis. He received the Novel Prize in Physiology or Medicine in 1903 for his contribution to treatment of diseases with "concentrated light radiation". Finsen designed the first carbon-arc lamp for phototherapy which was soon replaced by the less expensive mercury vapor lamp. By 1905, phototherapy was standard treatment for conditions such as eczematous dermatitis and psoriasis.[8] Sunlight phototherapy centers sprung up in the Western World; the favorite locations were Switzerland in Europe and the high mountains of Colorado in the United States where the beneficial "actinic" (UV) ray was plentiful due to the thin air. These centers claimed to treat a wide range of conditions: Skin conditions, chronic ulcers and wounds, cutaneous and other forms of tuberculosis, leukemia and "pseudoleukemia", and even skin cancer.[9] Unfortunately, scientific understanding of phototherapy lagged seriously behind the commercial successes. For example, an advocate for heliotherapy in Colorado have this to say about heliotherapy in 1915: "*The actinic (UV) rays falling on a white skin are nearly all reflected back and do not penetrate, but when the white skin becomes bronze and form deposit of pigments this pigment acts as a trans-former to the actinic rays and they then pass freely into the skin. The actinic rays now allowed to penetrate into or through the skin act on the corpuscles of the blood... stimulating and vitalizing them before they are carried by the general circulation to all organs and tissues of the body.*"[10]

The demise of phototherapy began with the availability of drug treatments for tuberculosis. Heliotherapy centers for tuberculosis were no longer necessary. The advent of antibiotics eventually replaced the function of phototherapy as treatment center for skin infections. Today, phototherapy is included as one of the treatment procedure for conditions such as psoriasis, acne vulgaris, eczema, and rickets. It is also an experimental treatment procedure for seasonal affective disorder, childhood failure to thrive, and jet lag.[11]

***Early warnings about UV radiation.*** For a long time, it was believed that the heat was responsible for tanning induced by exposure to the sun. In 1808, the German scientist Placidus Heinrich proposed that the light, not the heat of the sun, was responsible for tanning. In the 1840s and 1850, European scientists working with carbon arc lamps noticed that some components in the light caused eye problems and skin burn. It was only in 1885 that Paul Unna of Germany suggested that the violet end of the solar spectrum was responsible for the increased pigmentation of the skin.[12] Unna then provided a detail histopathological description of a pre-cancerous, degenerative condition of the skin commonly found in sailors exposed to too much sun.[13] In 1889, Widmark showed that it was the UV part of the spectrum from the arc lamps that were responsible for skin burns.[6] In 1906, the American dermatologist, James Hyde,[14] described a form of skin cancer that occurred in patients with xeroderma pigmentosa[b] and suggested that it shared a common characteristic with the cancerous changes that occurred on sailor's skin in that both are induced by excessive sunlight. A year later the French dermatologist William Dubreuith[15] studied epidemiological data on skin cancers and made the original observation that skin cancer of the face in seniors is "...*not solely caused by aging. It is most likely the result of chronic sun exposure... It affected mostly farmers and people who worked outside all their life under direct sun exposure, Light skinned people, whose skin contains less pigment, seem to be more affected than darker skinned people*". The UV-theory of skin cancer was supported by an animal experiment reported by G.M. Findlay[16] in 1928. He exposed depilated albino mice to UV light emitted from a mercury vapor lamp for a period of 8-months and observed pre-malignant and malignant growth on the skin. Despite mounting evidence, Michael Albert and Kristen Ostheimer[8] pointed out that excessive sunlight exposure as a cause of skin cancer was not fully embraced by the medical community. For example, an editor stated in a 1929 issue of the *Journal of the American Medical Association* that "...*there is no evidence available that exposure to the sun predisposes to epithelioma of the skin*".[17]

## Hazards of UV Radiation

*Mechanism of action of solar UV radiation damage.* In order to understand the biological effects of solar UV radiation, it is necessary to study the broad spectrum UV radiation (100–400 nm) as well as the separate components (UVA, UVB, UVC). Evidence from cellular, animal and epidemiological studies indicated that broad spectrum UV radiation causes DNA damage, suppression of the immune system, tumor promotion, and mutations in the p53 tumor-suppressor gene. UVA, UVB, and UVC as individual components of broad-spectrum

---

[b] Xeroderma pigmentosa (XP) is a condition inherited as an autosomal recessive trait and is characterized by photosensitivity, pigmentary changes, premature skin ageing and malignant tumour development. The underlying cause is the inherited defect in enzymes involved in the repair of damaged DNA such as those caused by sunlight.

UV radiation have been shown to cause genetic damage in *in vitro* test systems (bacteria, yeast, rodent cells, and human cells). There are indications in humans that, exposure to each of the three components of broad spectrum UV radiation causes DNA damages. UVA's biological effects are indirect and largely the result of energy transferred through reactive oxygen intermediates, whereas UVB and UVC are absorbed by DNA and directly damage DNA through base modifications.[18,19,52] The underlying basis of UV radiation-induced immunosuppression has not been fully elucidated but again damage to DNA has been proposed as a critical molecular event.[20]

The weight of evidence supports solar UV-B radiation as critical to the initiation of malignant melanoma. Chronic exposure to UV-A radiation may also play a contributory role in the progression of malignant melanoma. Exposure to solar UV-B radiation is believed to be the predominant environmental risk factor for both basal cell carcinoma and squamous cell carcinoma.

**Sunburn and Photoaging.** Excessive exposure to UV causes sunburn, which is noticeable 3 to 4 hours after exposure, peaking at around 24 hours. Long-term exposure to UV can lead to photoaging. UV-induced blistering sunburn in childhood and teen years is a risk factor for basal cell carcinoma and malignant melanoma, whereas excessive chronic cumulative exposure is a risk factor for squamous cell carcinoma.[58]

*Skin Cancers.* There are three main types of skin cancers in humans: cutaneous malignant melanoma, and the non-melanoma basal cell carcinoma and squamous cell carcinoma. Both the International Agency for Cancer Research (IARC)[21,53] and the U.S. Toxicological Program (US NTP)[52] have conducted thorough reviews on the carcinogenic effects of UV radiation. Although the classification approach and nomenclature differ slightly, the overall conclusion is similar — solar UV radiation and its separate components are either carcinogenic to humans or anticipated to be a human carcinogen. The classification results are summarized in Table 1.

Cancer of the skin is the most common of all cancers. One of every three cancers diagnosed worldwide is a skin cancer. In the

**Table 1.** Classification of Carcinogenicity of Solar Radiation and UVA, UVB and UVC.

| | US NTP (2011)[52] | IARC (1992, 2009)[21,53] |
| --- | --- | --- |
| Solar radiation | Known to be a human carcinogen | Carcinogenic to humans |
| Exposure to tanning devices | Known to be a human carcinogen | Carcinogenic to humans |
| Broad spectrum (UVA + UVB + UVC) | Known to be a human carcinogen | Carcinogenic to humans |
| UVA | Anticipated to be a human carcinogen | Probably carcinogenic to humans |
| UVB | Anticipated to be a human carcinogen | Probably carcinogenic to humans |
| UVC | Anticipated to be a human carcinogen | Probably carcinogenic to humans |

U.S. for example, more than 3.5 million cases of skin cancer are diagnosed each year. Basal and squamous cell cancer being the predominant types. Melanoma skin cancer accounts for less than 5% of the total but causes a great majority of skin cancer death.[56,57] The highest rates are found in countries where people are fairest-skinned and where the sun tanning culture is strongest: Australia, New Zealand, North America and Northern Europe. The annual incidence of melanoma varies geographically from between 4 and 24 per 100,000 in Europe and the United States to over 70 per 100,000 in higher ambient UV radiation regions of Australia and New Zealand (Fig. 1). In Australia, melanoma is currently the third most common cancer in men and women overall and the common-est in women aged 17–33 years of age.[22-24] In the U. S. the incidence rates for melanoma have been rising for at least 30 years.[57] The incidence of basal cell carcinoma and squamous cell carcinoma in many countries experience significant increase year after year.[22] For example, the incidence of basal cell carcinoma increased by 3% per year from 1996–2003 in the UK, and the incidence of squa-mous cell carcinoma increased four-fold from 1960–2004 in Sweden.[23]

The low incidence of skin cancers in darker skinned people[50] is primarily a result of photoprotection provided by increased epidermal melanin, which filters twice as much UV radiation as does the epidermis of Caucasians.[51] In the United States, for example, skin cancer represents approximately 20% to 30% of all neoplasm in Caucasians, 2% to 4% of all neoplasm in Asians, and 1% to 2% of all neoplasm in blacks and Asian Indians.[54] However, it should be noted that skin cancers in people with darker skin are often diagnosed at a later stage resulting in greater morbidity and mortality.[54]

*Eye diseases.* The eye and the skin are the two organs that are directly exposed to solar radiation. Acute exposure to solar UV radiation induces photokeratitis and photoconjunctivitis. High intensity exposure, such as looking directly at the sun, can cause severe retina burn. Despite well-publicized warning against direct viewing of a solar eclipse, a British study showed that increased incidence of solar retinopathy still occured days after the event.[54]

Chronic UV radiation is associated with the wide spread occurrence of cortical cataract. In 2004, 53.8 million people were estimated to have suffered moderate or severe disability resulting from cataract.[25] A pterygium is a benign growth of conjunctiva (the surface tissue of the white of the eye) that extends onto the cornea. Recent studies support an association between higher levels of sun exposure and the development of both primary and recurrent pterygium. Limited evidence indicates that there may be a link between solar UV-B radiation and the development of ocular melanoma.[23] Age-related macular degeneration is the most frequent cause of vision loss in humans living in developed countries. There is epidemiological and experimental evidence both for and against a role for exposure to UV radiation.[25] Common eye diseases resulting from chronic exposure to UV radiation, such as pterygium and cataract, create an immense social and financial burden globally.[22]

*Effect on the immune system.* In 1976, Kripke and Fisher[26] demonstrated that exposure of mice to UV radiation led to suppression of cell-mediated immunity within a short time following irradiation.

Since then more immune suppressive effects have been reported and active research is underway to elucidate the underlying mechanisms. The results from human and mouse studies so far provide strong evidence that UV-induced immune suppression plays a role in the development skin cancers. There is some evidence in humans that UV radiation suppresses the immune response to *herpes simplex virus* (HSV), which is responsible for cold sores, and *human papillomavirus* (HPV), which causes warts and in some instances, cancers of the cervix and penis. There is some indication that immune response to vaccines or the resistance to reinfection following vaccination may be affected by UVR exposure, but the evidence thus far is not definitive.[23,27]

***Adverse effects of sunbed use.*** The practice of indoor tanning is widespread in developed countries, and most popular in Northern Europe and the USA. Melanoma is the most common form of cancer among young adults aged 25–29 years and the second most common cancer in those aged 15–29 years. The International Agency for Research on Cancer (IARC)[21] has classified the use of UV-emitting tanning devices (sunbeds. solariums) as a human carcinogen. A major concern is the prominent and consistent increase in risk for melanoma and non-melanoma skin cancer in people who first used indoor tanning facilities in their twenties or teen years.[28,55] A melanoma outbreak in Iceland has been attributed to increased sunbed use.[29] Case-control studies conducted in Australia and the U.S. showed that frequent indoor tanning increased melanoma risk, regardless of age when indoor tanning began.[30,31,41] J-F Doré and M-C Chignol[55] reviewed the latest studies and noted the alarming statistics that up to 40–50% of teenagers in Northern Europe and the U.S.A had ever used indoor tanning. They recommend that use of sunbeds should be strongly discouraged, and banned for people under the age of 18.

## Benefits of UV Radiation

The primary benefit of solar radiation is the synthesis of vitamin D on the skin through the interaction of UV-B with 7-dehydrocholesterol.

The UNEP cautioned that the potential benefit of vitamin D, produced as a result of exposure to solar UV-B radiation, should be balanced against the harmful outcomes, such as skin cancers and cataract.[22]

Atopic dermatitis, psoriasis and a number of other skin conditions have been shown to be responsive to UV phototherapy.[32]

There are indications that solar UV radiation has some beneficial effect on autoimmune diseases such as type I diabetes mellitus, systemic lupus erythematosus, rheumatoid arthritis, and multiple sclerosis. It is suggested that UV-induced immunosuppression may be the basis of this beneficial effects.[33] An alternate explanation is that solar radiation promotes the synthesis of vitamin D which has some immunosuppressive activities.[34,35]

There are studies that suggest UV irradiation stimulates the production of several antimicrobial peptides which form part of the immune response of the skin. Other studies suggest that UVR exposure could be beneficial in at least some forms of asthma,[27] but more work is required to substantiate this view.

## Is Sunscreen Effective Against Skin Cancer?

The Sun Protection Factor (SPF) is a universally accepted numerical scale indicating the effectiveness of a sunscreen against sunburn. The method is based on determining the minimum erythematosus dose (MED), which is defined as the smallest amount of energy required for triggering erythema (redness of the skin) in areas of protected and unprotected skin. The SPF value is then calculated as the ratio between the MED of protected and unprotected skin.[36] The initial intention of sunscreens is, therefore, for protection of the skin against sunburn. As our understanding of the relationship between solar radiation and skin cancers grows, the focus of attention has been shifted to the effectiveness of sunscreens in the prevention of skin cancer. Routine use of sunscreen has been shown to be effective in reducing the risk of squamous cell carcinoma.[38,39] In addition, study by Green and co-workers[30] has shown that melanoma may also be prevented by regular sunscreen use in adults. Most countries are encouraging the use of "broad spectrum" sunscreens that protect not only UV-B but also the more penetrating UV-A.[42,43,44] However, there is concern that the information may be misunderstood by the public to mean that unrestricted exposure to the sun is safe when sunscreen is used.[40,41]

## Practical Points

- The World Health Organization[45] has a practical guide with detailed sun protection messages for the general public. It con-

**Table 2.** Popular Concept of UV Radiation Hazards.[45]

| False | True |
|---|---|
| A suntan is healthy | A tan results from your body defending itself against further damage from UV radiation |
| A tan protects you from the sun | A dark tan on white skin offers only limited protection equivalent to an SPF of about four |
| You cannot get sunburn on a cloudy day | Up to 80% of solar UV radiation can penetrate light cloud cover. Haze in the atmosphere can even increase UV radiation exposure |
| You cannot get sunburn while in the water | Water offers only minimal protection from UV radiation, and reflections from water can enhance your UV radiation exposure |
| UV radiation during the winter is not dangerous | UV radiation is generally lower during the winter months, but snow reflection can double your overall exposure, especially at high altitude. Pay particular attention in early spring when temperatures are low but the sun's rays are unexpectedly strong |
| Sunscreens protect me so I can sunbathe much longer | Sunscreens should not be used to increase sun exposure time but to increase protection during unavoidable exposure. The protection they afford depends critically on their correct application |
| If you take regular breaks during sunbathing you will not get sunburn | UV radiation exposure is cumulative during the day |
| If you do not feel the hot rays of the sun you will not get sunburn | Sunburn is caused by UV radiation which cannot be felt. The heating effect is caused by the sun's infrared radiation and not by UV radiation |

tains a true or false table (Table 2) which is particularly instructive for the general population.

- The American Academy of Pediatrics published a technical report on the hazard of UV-radiation on children and adolescent that included practical advices.[46]
- The International Commission on Non-Ionizing Radiation Protection[47] provided a list of people groups that are at particularly high risk of incurring adverse health effects from UVR, and therefore should be particularly counseled against the use of tanning appliances. The risk groups included individuals <18 years of age, individuals who have large numbers of nevi (moles), and individuals who tend to freckle.
- Both the U.S. Centers for Disease Control[48] and Prevention and the Cancer Council Australia[49] published guidelines to assist schools interested in UV risk reduction.
- Pay attention to the daily Global Solar UV Index (Fig. 1) which informs individuals about the strength of the sun's UV radiation.
- People with darker skin should also be aware of skin cancer because in this group, skin cancer is often present at an advanced stage, and thus, worse prognosis.

## References

1. Norval M, Lucas RM, Cullen AP *et al.* (2011) The human health effects of ozone depletion and interactions with climate change. *Photochem. Photobiol. Sci.* **10**: 199–225.
2. United Nations Environment Programme, Environmental Effects Assessment Panel. (2012) Environmental effects of ozone depletion and its interactions with climate change: Progress report, 2011. *Photochem. Photobiol. Sci.* **11**: 13–27.
3. Aucamp PJ. (2007) Questions and answers about the effects of the depletion of the ozone layer on humans and the environment. *Photochem. Photobiol. Sci.* **6**: 319–330.
4. Cullen AP. (2011) Ozone depletion and solar ultraviolet radiation: Ocular effects, a United Nations Environment Programme Perspective. *Eye Contact Lens* **37**: 185–190.
5. Psalms 74:16; 104:19. The Bible KJV.

6. Hockberger PE. (2002) A history of ultraviolet photobiology for humans, animals and microorganisms. *Photochem. Photobiol.* **76**: 561–579.

7. DeBroff BM, Pahk PJ. (2003) The ability of periorbitally applied antiglare products to improve contrast sensitivity in conditions of sunlight exposure. *Arch. Ophthalmol.* **121**: 997–1001.

8. Albert MR, Ostheimer KG. (2002) The evolution of current medical and popular attitudes toward ultraviolet light exposure: Part 1. *J. Am. Acad. Dermatol.* **47**: 930–937.

9. Lancet [editorial]. (1911) The sun in poisonous dose. *Lancet* **2**: 533.

10. Gardiner CF. (1915) Heliotherapy in Colorado. *Trans. Am. Climatol. Clin. Assoc.* **31**: 184–191.

11. Alberts JS. (2010) Sunshine: Clinical friend or foe? *Am. J. Med.* **123**: 291–292.

12. Henry W. Lim, Herbert Hönigsmann, and John L. M. Hawk (eds.). (2007) *Photodermatology*. Section 1. History and Basic Principles. Informa Healthcare, USA.

13. Unna PG. (1896) *The Histopathology of the Diseases of the Skin*. (Translated by N Walker) Wm. F. Clay, Edinburgh.

14. Hyde JN. (1906) On the influence of light in the production of cancer of the skin. *Am. J. Med. Sci.* **131**: 1–22.

15. Dubreuilh W. (1907) Épithéliomatose d'origine solaire. *Ann. Dermatol. Syphiligr. (Paris)* **8**: 387–416.

16. Findlay GM. (1928) Ultra-violet light and skin cancer. *Lancet* **2**: 1070–1073.

17. Ultraviolet and cancer [query to the editor]. (1929) *J. Am. Med. Assoc.* **93**: 1087.

18. Molho-Pessach V, Lotem M. (2001) Ultraviolet radiation and cutaneous carcinogenesis. *Curr Probl. Dermatol.* **35**: 14–27.

19. Young C. (2009) Solar ultraviolet radiation and skin cancer. *Occup. Med.* **59**: 82–88.

20. Schwarz T, Schwarz A. (2011) Molecular mechanisms of ultraviolet radiation-induced immunosuppression. *Eur. J. Cell Biol.* **90**: 560–564.

21. IARC. (2009) *IARC Monographs on the Evaluation of Carcinogenic Risks to Humans* Volume 100, A Review of Human Carcinogens Part D: Radiation. International Working Group on Cancer, Lyon, France.

22. United Nations Environment Programme, Environmental Effects Assessment Panel, Andrady AL, Aucamp PJ, Austin AT *et al.* (2012)

Environmental effects of ozone depletion and its interactions with climate change: progress report, 2011. *Photochem. Photobiol. Sci.* **11**: 13–27.

23. Norval M, Lucas RM, Cullen AP *et al.* (2011) The human health effects of ozone depletion and interactions with climate change. *Photochem. Photobiol. Sci.* **10**: 199–225.

24. *IARC Monographs on the Evaluation of Carcinogenic Risks to Humans.* (2009) Part D: Radiation Volume 100, International agency for Research on Cancer, Lyon, France.

25. Lucas RM. (2011) An epidemiological perspective of ultraviolet exposure–public health concerns. *Eye Contact Lens* **37**: 168–75.

26. Kripke ML, Fisher MS. (1976) Immunologic parameters of ultraviolet carcinogenesis. *J. Natl Cancer Inst.* **57**: 211–215.

27. Norval M, Halliday GM. (2011) The consequences of UV-induced immunosuppression for human health. *Photochem. Photobiol.* **87**: 965–977.

28. IARC. (2006) *Exposure to Artificial UV Radiation and Cancer*, Volume 1, IARC Working Group Report, International Working Group on Cancer, Lyon, France.

29. Hery C, Tryggvadottir L, Sigurdsson T *et al.* (2010) A melanoma epidemic in Iceland: Possible influence of sunbed use. *Am. J. Epidemiol.* **172**: 762–767.

30. Green AC, Williams GM, Logan V, Strutton GM *et al.* (2011) Reduced melanoma after regular sunscreen use: Randomized trial follow-up. *J. Clin. Oncol.* **29**: 257–263.

31. Lazovich D, Vogel RI, Berwick M *et al.* (2010) Indoor tanning and risk of melanoma: A case-control study in a highly exposed population. *Cancer Epidemiol. Biomarkers Prev.* **19**: 1557–1568.

32. Walker D, Jacobe H. (2011) Phototherapy in the age of biologics. *Semin. Cutan. Med. Surg.* **30**: 190–198.

33. Norval M, Halliday GM. (2011) The consequences of UV-induced immunosuppression for human health. *Photochem. Photobiol.* **87**: 965–77.

34. Kamen DL, Tangpricha V. (2010) Vitamin D and molecular actions on the immune system: Modulation of innate and autoimmunity. *J. Mol. Med. (Berl)* **8**: 441–450.

35. Szodoray P, Nakken B, Gaal J *et al.* (2008) The complex role of vitamin D in autoimmune diseases. *Scand. J. Immunol.* **68**: 261–269.

36. US Food and Drug Administration (FDA). Department of Health, Education and Welfare.

37. U. S Department of Health and Human Services. (1978) *Sunscreen Drug Products for Over-the-Counter Drugs: Proposed Safety, Effective and Labeling Conditions,* Vol. 43, No. 166, pp. 38206–38269. Federal Register, U. S. Government Printing Office.

38. Green A, Williams G, Neale R *et al.* (1999) Daily sunscreen application and betacarotene supplementation in prevention of basal-cell and squamous-cell carcinomas of the skin: A randomized controlled trial. *Lancet* **354**: 723–729.

39. van der Pols JC, Williams GM, Pandeya N *et al.* (2006) Prolonged prevention of squamous cell carcinoma of the skin by regular sunscreen use. *Cancer Epidemiol. Biomarkers Prev.* **15**: 2546–2548.

40. Autier P, Doré JF, Eggermont AM, Coebergh JW *et al.* (2011) Epidemiological evidence that UVA radiation is involved in the genesis of cutaneous melanoma. *Curr. Opin. Oncol.* **23**: 189–196.

41. Lazovich D, Vogel RI, Berwick M *et al.* (2011) Melanoma risk in relation to use of sunscreen or other sun protection methods. *Cancer Epidemiol. Biomarkers Prev.* **20**: 2583–2593.

42. US FDA. (2011) Guidance for Industry Enforcement Policy–OTC Sunscreen Drug Products Marketed Without an Approved Application. Drat Guidance. U.S. Department of Health and Human Services Food and Drug Administration Center for Drug Evaluation and Research (CDER).[http://www.fda.gov/downloads/Drugs/Guidance ComplianceRegulatoryInformation/Guidances/UCM259001.pdf].

43. Burnett ME, Wang Q. (2011) Current sunscreen controversies: A critical review. *Photodermatol. Photoimmunol. Photomed.* **27**: 58–67.

44. Fourtanier A, Moyal D, Seite S. (2012) UVA filters in sun-protection products: Regulatory and biological aspects. *Photochem. Photobiol. Sci.* **11**: 81–89.

45. WHO. (2002) A Global Solar UV Index, A Practical Guide. WHO/SDE/OEH/02.2. World Health Organization, Geneva. [http://www.unep.org/pdf/Solar_Index_Guide.pdf].

46. Balk SJ, Council on Environmental Health, Section on Dermatology. (2011) Ultraviolet radiation: A hazard to children and adolescents. *Pediatrics* **127**(3): 791–817.

47. IARC Working Group. (2006) The association of use of sunbeds with cutaneous malignant melanoma and other cancers: A systematic review. *Int. J. Cancer* **120**: 1116–1122.

48. U.S. Centers for Disease Control and Prevention. (2002) Guidelines for school programs to prevent skin cancer. *Morb. Mortal Wkly. Rep.* **51**(RR-4).

49. Cancer Council Australia. *UV Risk Reduction: A Planning Guide for Secondary School Communities*, The Cancer Council Australia. [http://www.cancer.org.au/File/Cancersmartlifestyle/UV%20risk% 20reduction.pdf].

50. Bradford PT. (2009) Skin cancer in skin of color. *Dermatol. Nurs.* **21**: 170–178.

51. Montagna W. (1991) The architecture of black and white skin. *J. Am. Acad. Dermatol.* **24**: 29–37.

52. Report on Carcinogens. Twelfth Edition. (2011) U.S. Department of Health and Human Services. Public Health Service. National Toxicology Program. *IARC Monographs on the Evaluation of Carcinogenic Risks to Humans. Solar and Ultraviolet radiation*, Vol. 55. International agency for Research on Cancer. Lyon, France.

53. Gloster HM Jr, Neal K. (2006) Skin cancer in skin of color. *J. Am. Acad. Dermatol.* **55**: 741–760.

54. Wong SCK, Eke T, Ziakas NG. (2001) Eclipse burns: a perspective study of solar retionopathy following the 1999 solar eclipse. *Lancet* **257**: 199–200.

55. Dorè J-F, Chignol M-C. (2012) Tanning salons and skin cancer. *Photochem. Photobiol. Sci.* **11**: 30–37.

56. American Cancer Society. What are the key statistics about basal and squamous cell skin cancers? http://www.cancer.org/Cancer/SkinCancer-BasalandSquamousCell/DetailedGuide/skin-cancer-basal-and-squamous-cell-key-statistics. Last revised:1/20/2012.

57. American Cancer Society. What are the key statistics about Melanoma? http://www.cancer.org/Cancer/SkinCancer-Melanoma/DetailedGuide/melanoma-skin-cancer-key-statistics. Last revised:1/11/2012.

58. Jou PC, Feldman RJ, Tomecki KJ. (2012) UV protection and sunscreens: What to tell patients. *Cleveland Clin J. Med.* **79**: 427–436.

# CHAPTER 14

# HEAT

*His going forth is from the end of the heaven, and his circuit unto the ends*
*of it: And there is nothing hid from the heat thereof.*

Psalms 19:6. The Bible (KJV)

The sun constantly radiates energy through space to the Earth. The radiated solar energy is predominantly in the form of electromagnetic waves of the near infrared, UV and visible range (Fig. 1, Chapter 11). Roughly half of the solar energy impinging on earth is reflected or absorbed by the atmosphere and the remaining half reaches the Earth's oceans and landmass. The Earth absorbs some of the energy and warms up, and reflects the rest of the energy back to the atmosphere in the far infrared form. The atmosphere is warmed by this reflected energy, and re-emits the heat upward to space and downward back towards Earth and warms it up further. Among the constituents of the atmosphere, clouds, water vapor and carbon dioxide are the most avid absorbers of the earth's energy. This phenomenon of trapping heat on Earth is called the greenhouse effect. When the level of carbon dioxide in the atmosphere is increased by human activities, more energy is absorbed and more heat is trapped on Earth, hence the effect of global warming. Some liken the accumulation of carbon dioxide to adding another blanket on earth to keep the heat in. The above represents a highly simplified explanation of the greenhouse effect and global warming. There is still much debate on the extent of contribution of greenhouse gases on global warming and on climate changes. There are some scientists who consider the present warming trend as a cyclical phenomenon. Still others question the reliability of the data and the models used to assess and predict changes. The

majority of experts accept that global warming is a fact. It was also predicted that, as a result of global warming, the Earth will experience
more and more extreme weather patterns. The acceptance of this view
was reflected by the award of the 2007 Nobel Peace Prize to the
Internal Panel on Climate Change (IPCC). In the first IPCC[1] report, it
was predicted that global average temperature will increase by 0.15°C
to 0.3°C per decade for the period of 1990–2005. The actual observed
value was 0.2°C per decade.[2] The 2011 IPCC report[3] indicated that
there has been (since 1950) an overall decrease in the number of cold
days and nights and an overall increase in the number of warm days
and nights, and predicted that the warming trend will continue for the
period under investigation (up to 2100), and that it is very likely that
the length, frequency and/or intensity of warm spells, or heat waves,
will increase. Furthermore, the hot extremes, which covered less than
1% of Eath's surface in the past (1951–1980), now covers almost 10%
of the land surface.[43] One of the major concerns about global warming
raised by IPCC[4] is the impact of heat waves on human health. This
concern is echoed by many scientists, physicians and public health
officials; some experts even consider climate change as the biggest
global health concern of the 21st century.[5]

## Heat-Related Illnesses[10–13]

Most people associated heat-related illnesses with specific situations
or activities, e.g. athletes participating in vigorous exercises, children kept in cars in hot weathers, or vacationers suffering from sun
strokes after a long day on the beach. What is less known is that
people get life-threatening heat-related illnesses during heat waves,
and right at home. Echoing the warning given by IPCC mentioned
above, health professionals have been expressing their concerns
about the increasingly frequent occurrence of heat waves and the
increasing reports of heatstrokes.[5–9]

   At the early stages, symptoms of heat-related illness can be mild
and subtle and liable to be viewed as general discomfort, or confused
with other disease conditions. However, the disease can rapidly
progress to serious and life-threatening conditions. It is therefore

important to clearly recognize the symptoms. Adverse effects caused by excessive heat include a spectrum of conditions with outcomes that progress from muscle cramp to life-threatening multiple organ failure. Although the illness is also called hyperthermia or pyrexia, the following terms are now in common use:

**Heat cramp.** Heat cramp is the mildest condition which occurs at the early stage, typically as a result of physical exertion, or exposure to warm temperature. The symptoms are sweating, fatigue, thirst and muscle cramp.

**Heat exhaustion.** Symptoms of heat exhaustion include weakness, malaise, nausea with or without vomiting, dizziness, weak and rapid pulse, muscle cramp and headache. Patients are often sweaty but the body temperature is normal or just slightly higher than normal. Their mental state is usually normal and they may not realize that the condition is heat-related.

**Heat stroke.** There are two types of heat stroke: (1) classical heat stroke brought on by prolonged exposure to heat, typically over a few days, e.g. during a heat wave and (2) exertional heat stroke occurring abruptly during physical exertion in a hot environment. Heat stroke, some times called sun stroke, is characterized by core body temperature above 40°C due to the loss of the body's ability to thermo-regulate. Another characteristic is that sweating is usually absent. Other symptoms include rapid and strong pulse, rapid breathing, muscle cramp, vomiting, headache, metal confusion, and unconsciousness. Complications of heat stroke include inflammatory responses and functional impairment of the heart, lungs, kidneys, liver, muscle and central nervous system. Disseminated intravascular coagulation, and abnormally low blood potassium and glucose may also occur. These complications have severe health consequences and are life-threatening.

## Heat Indices, Heat Wave and Urban Heat Islands

*Heat indices.* To most people, heat wave is a physical perception — exceptional warm days and nights that cause extreme discomfort. To medical practitioners and public health officials, heat wave is a period of time when the danger of heat stroke is the highest, a time to issue warning and to prepare for emergency. To weather experts and epidemiologists, a primary concern, however, is how to quantitatively and objectively measure the way body experiences heat, hence the need of a heat index. The most effective way for the body to dissipate heat is through sweating. Therefore an "effective temperature" must include a humidity factor because humidity directly affects the body's ability to sweat. Unfortunately, there is the problem of too many formulas for calculating the "effective temperature". In the U.S. a heat index is used which is calculated by a formula that included air temperature and relative humidity.[14] The term used in Canada is humidex[15] which is calculated with a different formula taking into account air temperature and dew point temperature. The International Standard Organization has developed the Wet Bulb Globe Temperature Index (WBGT).[16] Australia adopted the apparent temperature formula based on a mathematical model of an adult walking outdoors, in the shade.[17] A joint European project, EuroHEAT[18] has its own formula for calculating apparent temperature that also included air temperature and dew point temperature. Finally, the Universal Thermal Climate Index was proposed by the International Society of Biometerology.[19]

*Heat wave.* One of the original definition of a heat wave is three or more consecutive days during which the air temperature is >32.2°C.[20] Danella D'Ippoliti[18] and her colleague proposed a definition of a heat wave as a period of at least two days with (1) the maximum apparent temperature exceeding the 90th percentile of the monthly distribution or (2) the minimum apparent temperature exceeding the 90th percentile and the maximum apparent temperature exceeding the median monthly value. Yet another index, DD (degrees-days of "exceedance"),[21] has been proposed. The formula for the DD index takes into consideration the number of days with daily maximum temperature ($T_{max}$) above a certain local threshold ($T_{threshold}$) as well as the magnitude of the "exceedance".

As much as heat-related illness is a global phenomenon, it is important that there should be an agreement on a single calculation for a heat index for international use. Heat wave is more difficult to define because of the different durations, the changes in maximum and minimum temperatures, and the size and location of the affected area. Nevertheless, the scientific community should at least agree upon a set of parameters to define a heat wave. Otherwise, quantitative comparison between heat waves at different times and regions will be impossible,[18] and progress on the understanding of heat wave and its health impacts will be impeded. In the mean time, the important thing is for the public to observe their local heat index.

***Urban heat islands.*** It should be noted that indices such as humidex and heat index provide a quantitative measure of the heat a person experiences in a broad geographical region. It is not specific enough to warn us of areas or locations within a region that may be warmer than that indicated by the index. By experience, we all know that urban area is warmer than the suburb and countryside, and that certain locations within an urban area are particularly warm. Scientists have a term for this phenomena — urban heat islands — and have been working to understand their characteristics. The major causes for the formation of urban heat islands are (1) the large amount of heat generated from urban structures as they absorb and irradiate solar radiation and (2) heat generated from anthropological sources.[22,23] A characteristic of the phenomenon is that the intensity is stronger at night time than day time. It is hoped that the knowledge gain from research on urban heat islands can be applied to urban planning and landscaping, building construction and placement, to alleviate the effect of heat wave in densely populated area.

## History and Consequences of Heat Waves

In 24 BC Aelius Gallus, the Roman governor of Egypt conducted a military campaign in Arabia. His legion advanced in a harsh environment — the sun, the desert, drinking water of questionable quality — and many perished. They were also afflicted with an ailment that was unlike any of the common diseases because it attacked the head and killed the patient forthwith. For those who survived the first stage, the disease descended to the legs and cause dire injuries.[24] Abderrezak Bouchama[25] considered this as the earliest description of a heat stroke.

Because of the physical exertion demanded of soldiers, often under extreme temperature conditions, it is not surprising that the earliest work on heat stroke was conducted by investigators in the field of military medicine. The pathophysiology of 125 fatal cases of heatstroke was studied by N. Malamud and colleagues[26] in 1946 with a detailed description of the multi-organ damages.

***Heat waves in the U.S.*** One of the earliest comprehensive studies of heat stroke in civilians was conducted by Martin Austin,[27] who examined 1,000 patients admitted to hospitals in St. Louis, USA, during the three consecutive heat waves that occurred in the summers of 1952, 1953 and 1954. He concluded that the aged people and patients with cardiovascular diseases have the higher incidences of heatstroke, and that poor prognosis was associated with a body temperature of over 41°C, coma, uremia, hypotension and hyperkalemia (high blood potassium). His succinct observation of the role of humidity and air movement in the development of heatstroke still holds true today: "*At an environmental temperature of 30°C the amount of body heat that can be lost by convection, conduction and radiation is drastically reduced. With higher environmental temperature the only method of heat lost is by evaporation with conversion of sweat to water vapor. Evaporation is influenced by the environmental humidity and movement of air. The higher the relative humidity and the lower the air movement, the less efficient is the sweating mechanism. When the sweating mechanism for the heat loss fails, the syndrome of heatstroke results...*"

One of the first studies to focus on groups most vulnerable to the effect of heat waves was conducted by Schuman.[28] He analyzed data from the urban areas of the cities of New York and St. Louis, U.S.A., and identified several subgroups that are at substantially higher risk: persons over age 65; people with low income, poor housing or living in crowded conditions; patients with diabetes, hypertension, arteriosclerosis, cardiovascular diseases, or chronic respiratory diseases. The intense heat and drought that visited the

Midwestern United States throughout the summer of 1980 provide an opportunity for T. Stephen Jones and colleagues[29] to analyze the morbidity and mortality outcomes in the city of St. Louis and Kansas City. In the month of July alone, there was a 56.8% increase in deaths due to all causes in St. Louis. In Kansas City it was 65.2%. They demonstrated that the ratios of heatstroke rates were 3:1 for non-white *versus* white persons and 6:1 for low versus high socioeconomic status. They recommended that preventive measures in future heat waves be directed toward the urban poor, the elderly, and persons of other-than-white races. Besides the health effects, the heat and drought also caused an estimated economic loss of 16 billion in dollars in 1980 values.

As more and more statistics is being accumulated, it becomes clear that heat wave is one of the most devastating of all natural phenomena in terms of human illness and lives. For example, data from the US from 1979–2003 showed that more people in the country died from extreme heat than from hurricanes, lightning, tornadoes, floods, and earthquakes combined.[a] During 1999–2003, a total of 3,442 deaths resulted from exposure to extreme heat were reported in the U.S.[30] Australia is another country where historically more deaths occurred due to heat wave than to any other natural hazards.[31] It was calculated from mortality data that, between 1997 and 1999, there was an increase of 1% in mortality for every degree of maximum daily temperature exceeding 20°C. Susan Williams *et al.*[40] provided some striking statistics of the impact of heat waves on the citizens of Perth, Australia (1994–2008). With a 10°C increment in temperature above threshold (20°C–36°C), there was a 9.8% increase in daily mortality and a 10.2% increase in renal-related visits to the emergency department.

The physical display and destructive power of natural disasters such as hurricanes and floods certainly are impressive and capture our interest. During this time of global warming, heat wave as a silent but ruthless killer must not escape our attention.

---

[a] If the data were extended to include the year 2005, the statistic would have been different. Close to 1000 death has been attributed directly to Hurricane Katrina,[46] which wreak havoc on the U.S. Gulf Coast states that year.

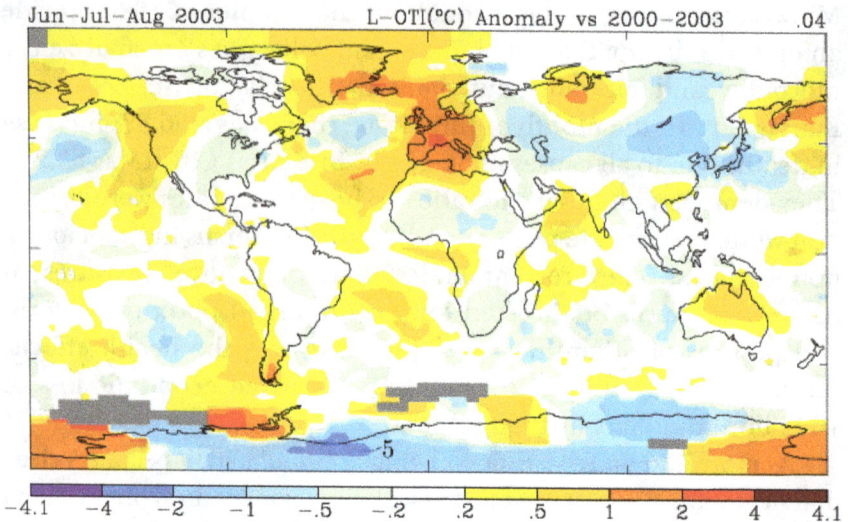

**Figure 1.** Surface air temperature anomalies in the summer (June–July) of 2003 with respect to 2000–2003 reference period. Data source: US National Aeronautic and Space Administration. Goddard Institute for Space Studies, GISS Surface Temperature Analysis.

***Heat waves in Europe.*** In the summer of 2003, the silent killer caught the attention of the world. This time the target was Europe. As clearly shown by the satellite image composed through surface temperature analysis, all of Western Europe was covered by a heat wave but France was the hardest hit (Fig. 1). It killed almost 15,000 people in France[32] and between 50,000 to 70,000 in all of Europe.[33]

During August 1–20, 2003 mortality rate rose by 60% when compared to the average values observed for the same period in 1999–2002, a rise that translated into 14,802[32] additional deaths. The regions in France with the highest heat-related mortality coincide with regions with the highest number of days with maximum temperatures above 35°C and minimum temperatures above 20°C.[34] Mathieu Oberlin and co-workers[35] depicted a picture of how unprepared health workers were in face of the surprise assault. They reviewed the medical record of patients aged 65 who were admitted to an emergency department during the heat wave and identified 42 patients with heat-related illness. None of the 42 patients were diagnosed correctly by attending physicians. As a result of lack of early diagnosis and proper treatment, 28.6% of these patients died in the emergency unit or after transfer to a hospital. They concluded that heat-related illness was under-estimated and under-diagnosed and recommended that, to improve the prognosis, emergency physician must be able to diagnose these patients early and begin cooling techniques. Benoît Misset and colleague[36] conducted a questionnaire survey of physicians heading intensive care units during the heat wave. Out of the 345 patients admitted to the intensive care unit with heat stroke, 216 died in the unit or in a hospital for a mortality rate of 62.6%. The median survival time was 13.1 days. This mortality rate was quite similar to that reported by Argaud *et al.*[32] of 58% and 71% after 28-days and 2-years, respectively. One of the practical but often neglected point raised by Misset[36] was that air conditioning in the intensive care unit had a beneficial effect in the treatment of patients with heat stroke. It was concluded from these studies that patients were at higher risk of suffering a heat stroke if they are aged, living alone, taking psychotropic or antihypertensive medications, suffering from respiratory, cardiovascular or nervous system diseases, or living in "depraved" cantons. These risk factors are consistent with those reported in earlier studies, some dating back to the 1950s.[27] It can only be said that France, and Europe in general, were caught off guard by the sudden appearance of the heat wave. The conclusion of Argaud[32] probably reflected the sentiment: *"Heatstroke is associated with poor outcomes in temperate urban areas. This could be explained at least in part by our lack of experience. Western temperate countries need to be more prepared for future heat waves"*.

Within two years of the heat wave, many European countries had drawn up more elaborate plans for the alert and management of heat waves.[33,41] At least 12 European countries have drawn up their own Heatwave Early Warning System.[42] The next heat wave came three years later during July 11–28, 2006, and resulted in 2065 heat-related death in France. This number was much lower than the predicted number of 6,452.[38] Although the nature of the two heat waves was not identical and prediction models have their uncertainties, the substantial drop in mortality rate suggested that France is better prepared this time.

***Mega-heat waves of 2010.*** In eastern Europe and large part of Russia, the exceptionally warm temperatures in the summer of 2010 caught the citizenry by surprise.[44] Preliminary estimate for Russia suggested a death toll of 55,000, an annual crop failure of ~25%, and US$15 billion of total economic loss (~1% domestic production). A large part of northern and eastern Asia also experienced extremely warm temperatures. Harbin, a city in northern China with a population of close to 10 million, saw a 41% jump in mortality in the period of June 7–11.[45]

Unlike many other environmental hazards and contaminants, the deadly effect of a heat wave is immediate and unequivocally apparent to all. Although we cannot abrogate the hazard, we can readily monitor it and, more importantly, we can greatly reduce the morbidity and mortality by tried-and-true preventive and mitigation measures. The real danger is that we may consider the last heat wave as an anomaly — just another "weather story" — rather than a fatal hazard that will return with certainty and in greater force.

## Practical Points[11,13,20,39]

Heat-induced illness is life-threatening. If you feel ill because of the high temperature, regardless of whether you are at rest or performing physical exercise, you must let other people know and see a health professional immediately.

Groups at greater risk of suffering from heat-related illness are:

— Infants and young children.
— People aged 65 or older.
— People living alone.
— People living in homes lacking air-conditioning.
— People with heart disease, high blood pressure, respiratory diseases, or mental disorders.
— People confined to beds or unable to care for themselves.

For an individual in normal health, the risk of contracting a heat-related illness during a heat wave is less if the following points and precautions are taken:

- Monitor daily weather reports, especially the heat index and pay attention to any heat alerts and advices.
- Drink more fluid.
- Avoid alcoholic drinks.
- Stay in cool and/or air-conditioned places whenever possible.
- Electric fans may help but are not effective when the temperature is >32°C.
- Take a cool shower whenever possible.
- Wear light weight, loose-fitting and light color clothing.

If you must be out in the heat, the following precaution should also be taken:

- Schedule outdoor activities during cooler times of the day.
- Reduce the level of physical activity and take frequent breaks.
- Drink plenty of fluid. Eat some salty food or take a sport beverage to replenish the salt and mineral you lose in sweats.
- Try to rest often in a shady area.
- Use the buddy system and check on each other.

## References

1. Intergovernmental Panel on Climate Changes. (1990) *Climate Change: The IPCC Impacts Assessment* (1990). In: McG Tegart WJ, Sheldon GW, Griffiths DC (eds.), report prepared for Intergovernmental Panel on Climate Change by Working Group II. Australian Government Publishing Service, Camberra, Australia.
2. Intergovernmental Panel on Climate Changes. (2007) The physical science basis. Contribution of working group I to the Fourth Assessment Report (AR$) of the Intergovernmental Panel on Climate Change. Solomon S, Qin D, Manning M, *et al.*, ed. Cambridge University Press.

3. IPCC. (2011) Intergovernmental Panel on Climate Change. Managing the Risks of Extreme Events and Disasters to Advance Climate Change Adaptation. A special report of Working Group I and Working Group II of the. [http://www.ipcc.ch/news_and_events/docs/ipcc34/ SREX_ FD_SPM_ final.pdf. November 18 2011 version].

4. *Contribution of Working Group II to the Fourth Assessment Report of the Intergovernmental Panel on Climate Change*, 2007. In Parry ML, Canziani OF, Palutikof JP *et al.* (eds.), Cambridge University Press.

5. Costello A, Abbas A, Allen A *et al.* (2009) Managing the health effects of climate change: Lancet and University College London Institute for Global Health Commission. *Lancet* **373**: 1693–1733.

6. Yoganathan D, Rom WN. (2001) Medical aspects of global warming. *Am. J. Ind. Med.* **40**: 199–210.

7. McGeehin MA, Mirabelli M. (2001) The potential impacts of climate variability and change on temperature-related morbidity and mortality in the United States. *Environ. Health Perspect.* **109** (Suppl 2): 185–189.

8. Kovats RS, Hajat S. (2008) Heat stress and public health: A critical review. *Annu. Rev. Public Health* **29**: 41–55.

9. O'Neill MS, Zanobetti A, Schwartz J. (2003) Modifiers of the temperature and mortality association in seven US cities. *Am. J. Epidemiol.* **157**: 1074–1082.

10. Bouchama A, Knochel JP. (2002) Heat stroke. *N. Engl. J. Med.* **346**: 1978–1988.

11. Yeo T. (2004) Heat stroke: A comprehensive review. *AACN Adv. Crit. Care* **15**: 280–293.

12. Epstein Y, Roberts WO. (2011) The pathopysiology of heat stroke: An integrative view of the final common pathway. *Scand. J. Med. Sci. Spor.* **21**: 742–748.

13. Centers for Disease Control and Prevention, Emergency Preparedness and Response, Extreme Heat: A prevention Guide to Promote your Personal Health and Safety. [http://www.bt.cdc.gov/disasters/extrem-eheat/heat_guide.asp].

14. US National Weather Service, National Oceanic and Atmospheric Administration. [http://firstgovsearch.gov/search?v%3Aproject=firstg ov&query=crh&affiliate=nws.noaa.gov].

15. Environment Canada. [http://www.ec.gc.ca/meteo-weather/default.asp?lang=En&n=86C0425B-1#h2].

16. Parsons K. (2006) Heat stress standard ISO 7243 and its global application. *Ind. Health* **44**: 368–379.

17. Steadman. RG. (1994) Norms of apparent temperature in Australia. *Aust. Met. Mag.* **43**: 1–16.

18. D'Ippoliti D, Michelozzi P, Marino C *et al.* (2010) The impact of heat waves on mortality in 9 European cities: Results from the EuroHEAT project. *Environ. Health* **9**: 37.

19. [UTCI. http://www.utci.org/isb.php].

20. Bouchama A, Knochel JP. (2220) Heat stroke. *N. Engl. J. Med.* **346**: 1978–1988.

21. García-Herrera R, Díaz J, Trigo RM *et al.* (2010) A Review of the European summer heat wave of 2003. *Crit. Rev. Environ. Sci. Technol.* **40**: 267–306.

22. Oke TR. (1988) The urban energy balance. *Prog. Phys. Geog.* **12**: 471–508.

23. Zhang K, Wang R, Shen C, Da L *et al.* (2010) Temporal and spatial characteristics of the urban heat island during rapid urbanization in Shanghai, China. *Environ. Monit. Assess.* **169**: 101–112.

24. Jarcho S. (1967) A roman experience with heat stroke in 24 B.C. *Bull. N. Y. Acad. Med.* **43**: 767–768.

25. Bouchama A. (1995) Heatstroke: A new look at an ancient disease. *Intensive Care Med.* **21**: 623–625.

26. Malamud N, Haymaker W, Custer RP. (1946) Heat stroke: A clinico-pathologic study of 125 fatal cases. *Mil. Surg.* **99**: 397–449.

27. Austin MG, Berry JW. (1956) Observations on one hundred cases of heatstroke. *J. Am. Med. Assoc.* **161**: 1525–1529.

28. Schuman SH. Fouillet A, Rey G. (1972) Patterns of urban heat-wave deaths and implications for prevention: Data from New York and St. Louis during July, 1966. *Environ. Res.* **5**: 59–75.

29. Jones TS, Liang AP, Kilbourne EM *et al.* (1982) Morbidity and mortality associated with the July 1980 heat wave in St Louis and Kansas City, Mo. *J. Am. Med. Assoc.* **247**: 3327–3331.

30. MMWR. (2000) Heat-related illnesses, deaths and risk factors—Cincinnati and Dayton, Ohio, 1999 and United States, 1979–1997. *Morb. Mortal Wkly. Rep.* **49**: 470–473.

31. Vaneckova P, Beggs PJ, de Dear RJ, McCracken KW *et al.* (2008) Effect of temperature on mortality during the six warmer months in Sydney, Australia, between 1993 and 2004. *Environ. Res.* **108**: 361–369.

32. Argaud L, Ferry T, Le Q-H *et al.* (2007) Short-and long-term outcomes of heatstroke following the 2003 heat wave in Lyon, France. *Arch. Intern. Med.* **167**: 2177–2183.

33. Montero JC, Mirón IJ, Criado JJ *et al.* (2010) Comparison between two methods of defining heat waves: A retrospective study in Castile-La Mancha (Spain). *Sci. Total Environ.* **408**: 1544–1550.

34. Garcia-Herrera R, Diaz J, Trigo RM *et al.* (2010) A review of the European summer heat wave of 2003. *Crit. Rev. Environ. Sci. Technol.* **40**: 267–306.

35. Oberlin M, Tubery M, Cances-Lauwers V *et al.* (2010) Heat-related illnesses during the 2003 heat wave in an emergency service. *Emerg. Med. J* **27**: 297–299.

36. Misset B, De Jonghe B, Bastuji-Garin S *et al.* (2006) Mortality of patients with heatstroke admitted to intensive care units during the 2003 heat wave in France: A national multiple-center risk-factor study. *Critical Care Med.* **34**: 1087–1092.

37. Fouillet A, Rey G, Laurent F *et al.* (2006) Excess mortality related to the August 2003 heat wave in France. *Int. Arch. Occup. Environ. Health* **80**: 16–24.

38. Fouillet A, Rey G, Wagner V *et al.* (2008) Has the impact of heat waves on mortality changed in France since the European heat wave of summer 2003? A study of the 2006 heat wave. *Int. J. Epidemiol.* **37**: 309–317.

39. Mattis JG, Yates AM. (2011) Heat stroke: Helping patients keep their cool. *Nurse Pract.* **36**: 48–52.

40. Williane S, Nitschke M, Weinstein P *et al.* (2012) The impact of summar temperature and heatwaves on mortality and mobidity in Perth, Australia 1999–2008, *Environ. Int.* **40**: 33–38.

41. Koppe C, Becker P. Comparison of operational heat health warning systems in Europe. Working Document of the Project "Improving Public Health Responses to Extreme Weather/Heat-waves-EuroHeat"; WHO Regional Office for Europe: Coperhagen, Denmark, 2007.

42. Lowe D, Ebi K, Forsberg B. (2011) Heatwave early warming systems and adaptation advice to reduce human health consequences of heat-waves. *In J. Environ. Res. Public Health* 8: 4623–4628.

43. Hansen J, Sato M, Ruedy R. (2012) Perception of climate change. *Proc Natl Acad Sci U S A*. In press.

44. Barriopedro D, Fischer EM, Luterbacher J, *et al.* (2011) The Hot Summer of 2010: Redrawing the Temperature Record Map of Europe. *Science* **332**: 220–224.

45. Lan L, Cui G, Yang C, *et al.* (2012) Increased Mortality During the 2010 Heat Wave in Harbin, China. EcoHealth. www.ecohealth.net.

46. Brunkard J, Namulanda G, Ratard R. (2008) Hurricane Katrina deaths, Louisiana, 2005. *Disaster Med Public Health Prep.* **2**: 215–223.

# NOTE

## Fukushima fallout and low dose effects

Radioactive fallout from Fukushima, mainly in the form of iodine 131 and cesium 134 and 137, was detected at very low level in North America within a week of the accident.[1] The potential health effect of the fallout is, understandably, of immediate interest to the general public. The health effect data will also be of great help in understanding the consequences of exposure to low level environmental radiation. Because the total amount of radioactive materials released from the Fukushima accident was much smaller than that from the Chernobyl explosion, it has been suggested that the health effect of the Fukushima fallout will be less and hence even more difficult to ascertain.[2,3] On the other hand, Mangano and Sherman studied selected morality data collated by the U.S. Centers for Disease Control and Prevention (CDC) and concluded that there was an excess of total deaths and infant deaths in the United States in the 14 weeks following the arrival of the fallout.[4] A projected 13,983 excess death, later revised to 21,851 deaths,[8] were mentioned. The conclusion was promptly challenged by three different authors.[5-7] One of the criticisms is that, due to a lack of understanding of the mechanisms of action, the reported correlation between radioactive fallout and excess mortality cannot prove a cause-effect relationship. Another concern is that the number of cities included in the study for the before and after Fukushima comparison was different. The choice of biostatistics used for the analysis of population data was also questioned. Thus, a trend analysis of the same CDC dataset yielded no significant increase in infant death after Fukushima.[7] These criticisms were countered immediately by Mangano and Sherman.[8] The debate illustrates the frustrating fact that 26 years after Chernobyl and 67 years after Hiroshima and Nagasaki, we still do not know what the level of risk is following radioactive fallouts or low dose exposures.[3]

## References

1. Thakur P, Ballard S, Nelson R. (2012) Radioactive fallout in the United States due to the Fukushima nuclear plant accident. *J. Environ. Monitor.* **14**: 1317–1324.

2. McLaughlin PD, Jones B, Maher MM. (2012) An update on radioactive release and exposures after the Fukushima Dai-ichi nuclear disaster. *Brit. J. Radiol.* **85**: 1222–1225.

3. Boice JD Jr. (2012) Radiation epidemiology: A perspective on Fukushima. *J. Radiol. Prot.* **32**: N33–N40.

4. Mangano JJ, Sherman JD. (2012) An unexpected mortality increase in the United States follows arrival of the radioactive plume from Fukushima: Is there a correlation? *Int. J. Health Services* **42**: 47–64.

5. Wolf A. (2012). Response to "An unexpected mortality increase in the United States follows arrival of the radioactive plume from Fukushima: Is there a correlation?" *Int. J. Health Services* **42**: 549–551.

6. Gale RP. (2012). Response to "An unexpected mortality increase in the United States follows arrival of the radioactive plume from Fukushima: Is there a correlation?" *Int. J. Health Services* **42**: 557–559.

7. Köblein A. (2012). Response to "An unexpected mortality increase in the United States follows arrival of the radioactive plume from Fukushima: Is there a correlation?" *Int. J. Health Services* **42**: 553–555.

8. Mangano JJ, Sherman JD. Fukushima update: Radioactive fallout and mortality increases in the United States: Is there a correlation? *Int. J. Health Services* **42**: 561–570.

# INDEX

www.ingramcontent.com/pod-product-compliance
Lightning Source LLC
Chambersburg PA
CBHW050549190326
41458CB00007B/1979